2.99

TECHNIQUES IN VISIBLE AND
ULTRAVIOLET SPECTROMETRY

# VOLUME FOUR
# UV SPECTROSCOPY

# UV Spectroscopy
# Techniques, instrumentation, data handling

## UV SPECTROMETRY GROUP

Edited by

**B. J. CLARK,**
*Pharmaceutical Chemistry,*
*University of Bradford*

**T. FROST**
*The Wellcome Foundation Ltd,*
*Temple Hill, Dartford, Kent*

and

**M. A. RUSSELL,**
*Merck Ltd,*
*Poole, Dorset*

**CHAPMAN & HALL**
London · Glasgow · New York · Tokyo · Melbourne · Madras

**Published by Chapman & Hall, 2–6 Boundary Row, London SE1 8HN**

Chapman & Hall, 2–6 Boundary Row, London SE1 8HN, UK

Blackie Academic & Professional, Wester Cleddens Road, Bishopbriggs, Glasgow G64 2NZ, UK

Chapman & Hall Inc., 29 West 35th Street, New York NY10001, USA

Chapman & Hall Japan, Thomson Publishing Japan, Hirakawacho Nemoto Building, 6F, 1-7-11 Hirakawa-cho, Chiyoda-ku, Tokyo 102, Japan

Chapman & Hall Australia, Thomas Nelson Australia, 102 Dodds Street, South Melbourne, Victoria 3205, Australia

Chapman & Hall India, R. Seshadri, 32 Second Main Road, CIT East, Madras 600 035, India

First edition 1993

© 1993 UV Spectrometry Group

Typeset in 10/12 Times by Expo Holdings, Malaysia
Printed in Great Britain by Richard Clay Ltd, Bungay, Suffolk

ISBN 0 412 40530 X

Apart from any fair dealing for the purposes of research or private study, or criticism or review, as permitted under the UK Copyright Designs and Patents Act, 1988, this publication may not be reproduced, stored, or transmitted, in any form or by any means, without the prior permission in writing of the publishers, or in the case of reprographic reproduction only in accordance with the terms of the licences issued by the Copyright Licensing Agency in the UK, or in accordance with the terms of licences issued by the appropriate Reproduction Rights Organization outside the UK. Enquiries concerning reproduction outside the terms stated here should be sent to the publishers at the London address printed on this page.

The publisher makes no representation, express or implied, with regard to the accuracy of the information contained in this book and cannot accept any legal responsibility or liability for any errors or omissions that may be made.

A catalogue record for this book is available from the British Library

Library of Congress Cataloging-in-Publication data available

# Contents

| | | |
|---|---|---|
| **Contributors** | | ix |
| **Preface** | | xi |
| **1** | **Principles of spectrophotometric measurements with particular reference to the UV-visible region**  *D. Thorburn Burns* | 1 |
| | 1.1 The nature of radiant energy and its interaction with matter | 1 |
| | 1.2 Frequency, wavelength and wavenumber | 3 |
| | 1.3 Units for wavelength | 5 |
| | 1.4 Origin and nature of spectra | 6 |
| | 1.5 Laws of absorption of light | 6 |
| | 1.6 Quality of a spectrum | 10 |
| | 1.7 Accuracy and precision of results | 12 |
| | 1.8 Production of a valid spectrum | 14 |
| | 1.9 Standards | 15 |
| | 1.10 Summary | 15 |
| | References | 15 |
| | Further reading | 15 |
| **2** | **Instrumentation for UV-visible and fluorescence spectroscopy**  *J.N. Miller* | 17 |
| | 2.1 Introduction | 17 |
| | 2.2 Instrumentation for absorption spectroscopy | 17 |
| | 2.3 Instrument components in absorption spectroscopy | 18 |
| | 2.4 Instrumentation for luminescence spectroscopy | 23 |
| | 2.5 Origin of molecular luminescence | 24 |
| | 2.6 Fluorescence intensity | 26 |
| | 2.7 Factors affecting fluorescence intensity | 28 |
| | 2.8 Fluorescence polarization | 29 |
| | 2.9 Instrumentation in fluorimetry | 30 |
| | 2.10 Fluorescence spectra and their correction | 32 |
| | 2.11 General methods in fluorescence spectroscopy | 34 |
| | Further reading | 34 |
| **3** | **Standards in UV spectroscopy**  *M.A. Russell* | 35 |
| | 3.1 Introduction | 35 |

vi    *Contents*

|  |  |  |
|---|---|---|
| 3.2 | Wavelength standards | 37 |
| 3.3 | Absorbance standards | 42 |
| 3.4 | Stray light | 47 |
|  | References | 51 |

**4    Spectroscopic multicomponent analysis**
*R.L. Tranter and B. Davies*                                                      52

|  |  |  |
|---|---|---|
| 4.1 | Introduction | 52 |
| 4.2 | Multivariate calibration techniques | 53 |
| 4.3 | Principles | 53 |
| 4.4 | Simple linear regression: a tutorial | 55 |
| 4.5 | The Kalman filter | 58 |
| 4.6 | Multicomponant calibration | 60 |
| 4.7 | Principal componont analysis and regression | 61 |
| 4.8 | Summary | 62 |
|  | References | 62 |

**5    UV-visible spectroscopic libraries**    *B. Davies*    63

|  |  |  |
|---|---|---|
| 5.1 | Introduction | 63 |
| 5.2 | Creating a spectroscopic database | 63 |
| 5.3 | Spectroscopic validity | 64 |
| 5.4 | Data storage | 65 |
| 5.5 | Techniques for searching spectral libraries | 66 |
| 5.6 | Spectral scaling | 68 |
| 5.7 | Comparison techniques | 69 |
| 5.8 | Comparison algorithms | 72 |
| 5.9 | Comparison metrics applied to a demonstration library | 74 |
|  | References | 77 |

**6    Spectra-structure correlation**    *T.L. Threlfall*    78

|  |  |  |
|---|---|---|
| 6.1 | Origin of spectra | 78 |
| 6.2 | Common absorption bands | 79 |
| 6.3 | Summary | 84 |
|  | References | 85 |
|  | Further reading | 86 |

**7    Colour**    *T.L. Threlfall*    88

|  |  |  |
|---|---|---|
| 7.1 | Colour perception | 88 |
| 7.2 | Colour vision theories | 90 |
| 7.3 | Colour measurement systems | 90 |
| 7.4 | Problems in colour measurement | 94 |
| 7.5 | Summary | 95 |
|  | References | 95 |

| | | |
|---|---|---|
| **8** | **Liquid chromatographic detection for multi-component analysis**   *B.J. Clark* | 97 |
| | 8.1   Introduction | 97 |
| | 8.2   High performance liquid chromatography | 99 |
| | 8.3   Hybrid detectors for HPLC | 103 |
| | 8.4   Rapid scanning spectroscopic detection | 107 |
| | 8.5   Conclusions | 118 |
| | References | 119 |
| **9** | **Assessing the validity of spectroscopic information**   *C. Burgess* | 121 |
| | 9.1   Concepts of validation | 121 |
| | 9.2   Stage 1: Fundamental | 121 |
| | 9.3   Stage 2: System control | 123 |
| | 9.4   Stage 3: Transformation | 123 |
| | 9.5   Stage 4: Interpretation | 125 |
| | 9.6   Summary | 125 |
| **10** | **Practical experiments**   *T. Frost and R.L. Tranter* | 126 |
| | 10.1   Introduction | 126 |
| | 10.2   The experiments | 126 |
| | Session 1: Fundamental principles | 127 |
| | Session 2: Collection of good quality spectra | 133 |
| | Session 3: Quantitative aspects | 136 |
| | Session 4: Multicomponent calculations | 139 |
| | References | 142 |
| **Index** | | 143 |

# Contributors

All the contributors are long-standing members of the UVSG; indeed five are former chairmen and a sixth currently holds that post. Their present affiliations are as follows:

Dr C. Burgess, Glaxo Manufacturing Services, Barnard Castle, Co Durham
Prof. D. Thorburn Burns, The Queen's University, Belfast
Dr B.J. Clark, University of Bradford, Bradford, West Yorkshire
Dr B. Davies, Glaxo Manufacturing Services, Barnard Castle, Co Durham
Mr T. Frost, The Wellcome Foundation Ltd, Dartford, Kent
Prof. J.N. Miller, University of Loughborough, Loughborough, Leicestershire
Dr M.A. Russell, Merck Ltd, Poole, Dorset
Dr T.L. Threlfall, York University, York, West Yorkshire
Dr R.L. Tranter, Glaxo Manufacturing Services, Barnard Castle, Co Durham

# Preface

The material in this book has been derived from a UV Spectroscopy Summer School organized by the Ultraviolet Spectrometry Group (UVSG) at the Burn Hall Centre near York in July 1990. Training and education in all matters relating to UV and Visible Spectrometry has always been a major function of the UVSG which consequently has a long history of organizing such Summer Schools. The first course was run in 1949, only one year after the Group was founded, and they have been a regular feature of the Group's activities ever since. The event is the only major training course in UV Spectroscopy currently run in the UK. During the 1980s a series of successful courses were organized at Loughborough University.

Recently there has been a renaissance in UV Spectroscopy with many exciting new techniques, instruments and data processing methods becoming available. These advances have primarily revolved around the developments in computer technology which have allowed the introduction of sophisticated instruments and software.

For the 1990 course the Group incorporated many of these developments and thoroughly revised the content of the course. The aim was to have a course that started with the most fundamental principles and progressed through to encompass advanced chemometric techniques. Much of the emphasis throughout was placed on the practical application of UV-visible spectroscopy. This book has the same aim and we hope it will be of use to anyone involved in using UV Spectroscopy in the widest sense.

Chapter 1 introduces the fundamental principles of the interaction of light and matter, with Chapter 2 discussing instrumentation both for UV-visible absorbance and fluorometric methods. Chapter 3 explores the standards that are available to check instrument performance and to ensure that correct results are obtained. The next two chapters introduce some of the more mathematical techniques for multicomponent analysis and also discusses spectral libraries. These are followed by a chapter on spectra and structure correlation and one on colour. Chapter 8 introduces some of the techniques used in the use of UV detection in HPLC and the final chapter looks at the important topic of validation and use of data.

Any of the topics discussed in these chapters could, and some have been, the subject of a book in their own right. This book therefore only seeks to act as an introduction. References are given to aid those who require further information.

The book concludes with a series of practical exercises which complement the lecture material and further illustrate the points raised. These exercises are intended for use in training scientists and technicians, and are flexible enough to allow easy modification to suit particular circumstances. However, such modifications must always be checked by experienced workers before being used for teaching purposes and are designed to be run under the supervision of a qualified analytical chemist. We hope the exercises will be of use both in industry and in academic environments.

Much work went into the production of the material in this book and proceeds from the sale of the book will go to the UVSG for the furtherance of UV spectroscopy. On behalf of the group our thanks go to all the contributors who have given their time outside their normal day to day work.

# 1 Principles of spectrophotometric measurements with particular reference to the UV-visible region

## 1.1 The nature of radiant energy and its interaction with matter

Radiant energy which is visible to the human eye is called light and covers the spectral region from about 400 to 750 nm (Fig. 1.1). The apparent colour of a material i.e. the light reflected by it, is the complement of the colour of the incident light absorbed (Table 1.1). For example, an aqueous solution of copper (II) sulphate appears blue because it absorbs yellow light. Colour and related matters are dealt with in more detail in Chapter 7. Investigations into the properties of light which were subsequently generalized to cover all spectral regions, reveal that it can behave in two different ways, that is there are two

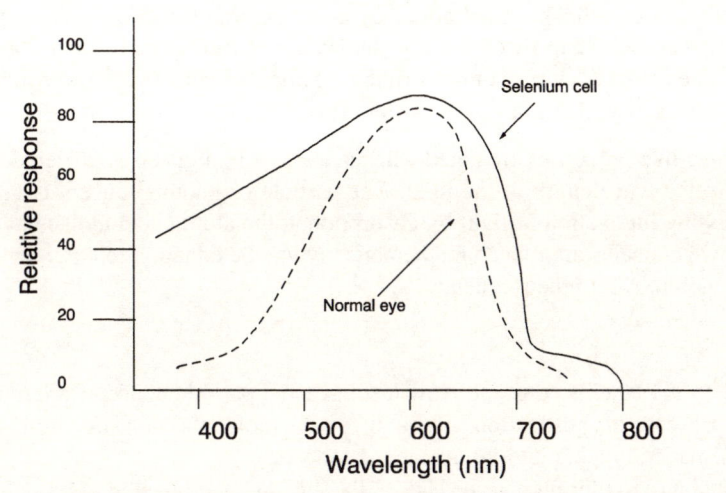

Fig. 1.1 *The spectrum of 'visible' light as seen by the eye compared with that detected by a selenium barrier layer cell with a protective glass cover.*

2   *Principles of spectrophotometric measurements*

**Table 1.1**  *Colours of visible radiation and their complements*

| Approximate wavelength range (nm) | Colour of reflected light | Colour of absorbed light |
|---|---|---|
| 400–465 | Violet | Yellow-green |
| 465–482 | Blue | Yellow |
| 482–487 | Greenish-blue | Orange |
| 487–493 | Blue-green | Red-orange |
| 493–498 | Bluish-green | Red |
| 498–530 | Green | Red-purple |
| 530–559 | Yellowish-green | Reddish-purple |
| 559–571 | Yellow-green | Purple |
| 571–576 | Greenish-yellow | Violet |
| 576–580 | Yellow | Blue |
| 580–587 | Yellowish-orange | Blue |
| 587–597 | Orange | Greenish-blue |
| 597–617 | Reddish-orange | Blue-green |
| 617–780 | Red | Blue-green |

different models for its behaviour dependent on the experimental arrangements. Light can act both with wave-like and particle-like properties. This gives rise to the so-called 'wave-particle duality'.

The nature of light always was, and remains, a difficult concept. For instance, Robert Boyle wrote in 1664 [1]:

> ...to give a perfect account of the theory of Vision and colours ... I would first know what Light is, and if, if it be a Body (as a Body or the Motion of a Body it seems to be) what kind of Corpuscles for size and shape it consists of, with what Swiftness they move Forwards and Whirl about their own Centres. Then I would know the Nature of Refraction, which I take to be one of the Abstrucest things (not to explicate Plausibly, but to explicate Satisfactorily) that I have met with in Physics...

Wave-like properties are noted when a beam of light meets a barrier similar in size to the wavelength of the light. The particle or photon concept is useful in discussing interaction of light by absorption at the atomic and molecular levels. The two concepts are related mathematically via the equation relating the energy of a photon, $\Delta E$, to the frequency, $\nu$,

$$\Delta E = h\nu \tag{1}$$

where $h$ is Planck's constant. How the wave and particle concepts are related in the process of absorption of light by a molecule can be shown non-mathematically in a pictorial manner as follows.

Light moving through space is described in classical physics terms as '*a force field in space with a characteristic wavelength, velocity and intensity*', (Fig. 1.2). A photon is thus considered as moving through space as a changing magnetic

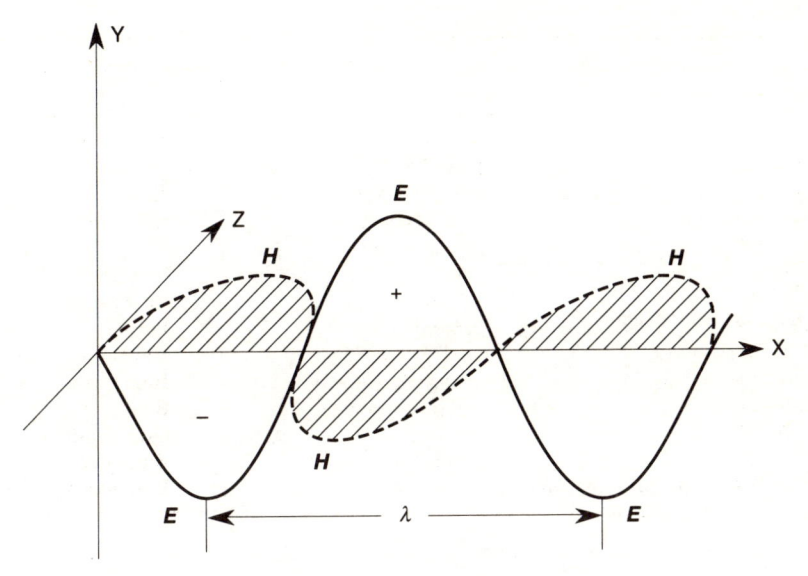

Fig. 1.2  *A plane polarized electromagnetic wave of single frequency and of wavelength, $\lambda$.  **E** – electric field vector; **H** – magnetic intensity vector. Reproduced from Comprehensive Analytical Chemistry, Volume XIX, Nowicka-Jahkowska, Gorczynska, Michalik and Wieteska (1986) with the permission of Elsevier.*

and electronic field. The electronic field can interact with electrons in matter and be absorbed.

In the case of an atom with a single electron in an $s$ orbital the encounter between the changing electrical field associated with the photon and the electron in the $s$ orbital may be envisaged by a change in direction and velocity of the electron, hence in its angular momentum. The energy absorbed, the photon or the light quantum thus promotes the orbital electron from the $s$ into the $p_z$ orbital, shown diagrammatically in Fig. 1.3.

The process of absorption of a photon by a molecule may be illustrated using ethene (ethylene) as a simple example. The encounter between the changing electronic field and the $\pi$ orbital may be pictured as the inversion of one $p_z$ orbital thus forming a $\pi^*$ antibonding molecular orbital, as illustrated below (Fig. 1.4).

## 1.2  Frequency, wavelength and wavenumber

Although, for ease of pictorial representation, the wavelength aspect of light was shown in Fig. 1.3 it should be noted that *frequency* is the distinguishing characteristic of light. Frequency, $v$, is related to wavelength, $\lambda$ and velocity, $V$ by the expression

$$v = \frac{V}{\lambda} \qquad (2)$$

4   Principles of spectrophotometric measurements

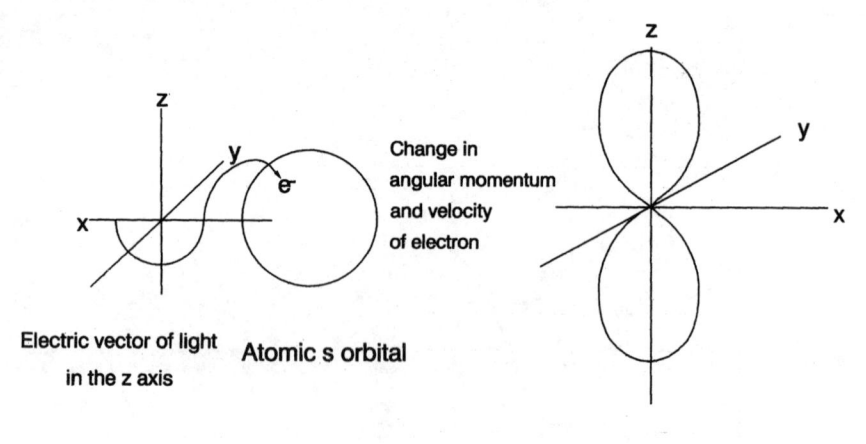

Fig. 1.3  Pictorial representation of the interaction of the electric vector for light with an electron in an s-atomic orbital.

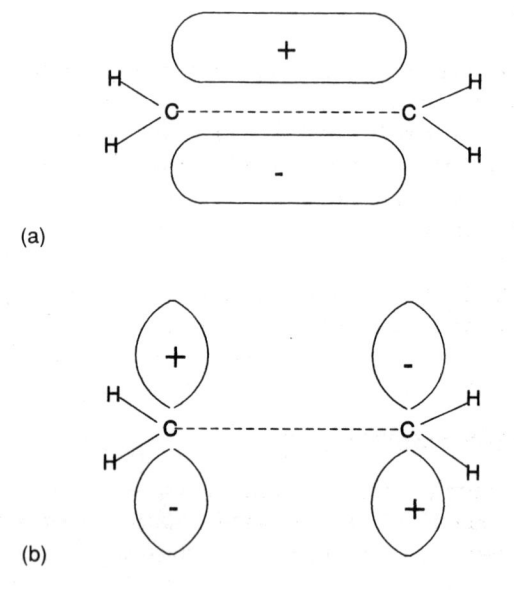

Fig. 1.4  Side views of (a) $\pi$ and (b) $\pi^*$ orbitals of ethene. The positive and negative signs represent the signs of the wavefunction.

Both velocity and wavelength depend on the medium through which the light wave travels (Fig. 1.5). For two media A and B,

$$v = \frac{V_A}{\lambda_A} = \frac{V_B}{\lambda_B} \qquad (3)$$

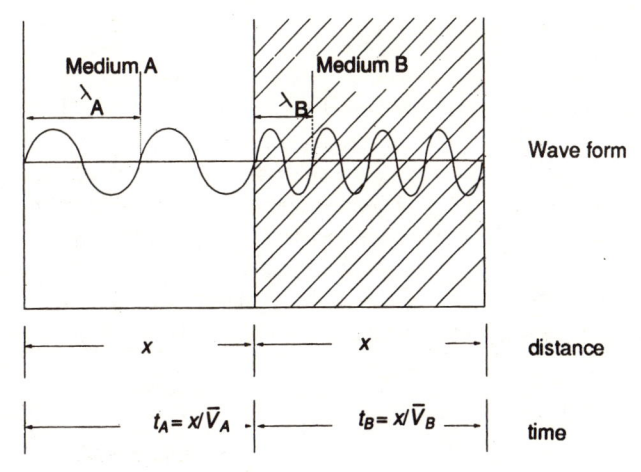

Fig. 1.5 *Effect of refractive index of matter upon wavelength, $\lambda$, and velocity, $\bar{V}$, of plane polarized light of frequency, $v$. Medium B has the higher refractive index.*

The change in velocity with change in medium gives rise to the phenomenon of refraction. In many experiments with a consistent medium throughout, velocity is not a variable and data are often reported in terms of $1/\lambda$ or wavenumber $\bar{v}$. The use of the symbols $v$ (frequency) and $\bar{v}$ (wavenumber) is potentially confusing. Spectra, i.e. the variation of some parameter such as the percentage transmission of light by a sample, with size of photon $\Delta E$, can be plotted on either scale depending on choice or custom in a particular spectral region. It should be noted that frequency and wavenumber are directly proportional to $\Delta E$ whereas $\lambda$ is inversely related to $\Delta E$.

## 1.3 Units for wavelength

Wavelength in the UV-visible spectral region (200–800 nm) may be expressed in terms of nanometers (nm) ångströms (Å) or millimicrons (m$\mu$), and are interrelated thus:

$$\begin{aligned} 1 \text{ nm} &= 10^{-9} \text{ m} \\ &= 10^{-7} \text{ cm} \\ &= 10 \text{ Å} \\ &= 1 \text{ m}\mu \\ 1 \text{ }\mu\text{m} &= 10^{-6} \text{ m} \end{aligned}$$

It is now customary to use nanometer and to avoid entirely the pre-SI ångström unit.

## 6 Principles of spectrophotometric measurements

## 1.4 Origin and nature of spectra

The energies of the photons in the region 200–800 nm permit excitation of outer valence electrons and inner shell, d–d transitions with associated vibrational levels (Fig. 1.6). When molecules are close together, as they normally are in spectrophotometric measurements on solutions, they exert influences on each other's energy levels which become broadened. Hence the vibrational fine structure apparent in the vapour phase is wholly or partly lost and the spectra appear as bands, as illustrated by the vapour phase and solution spectra of benzene (Fig. 1.7). The degree of loss of fine structure is, in many cases, dependent on the solvent and on the nature of the solute-solvent interactions.

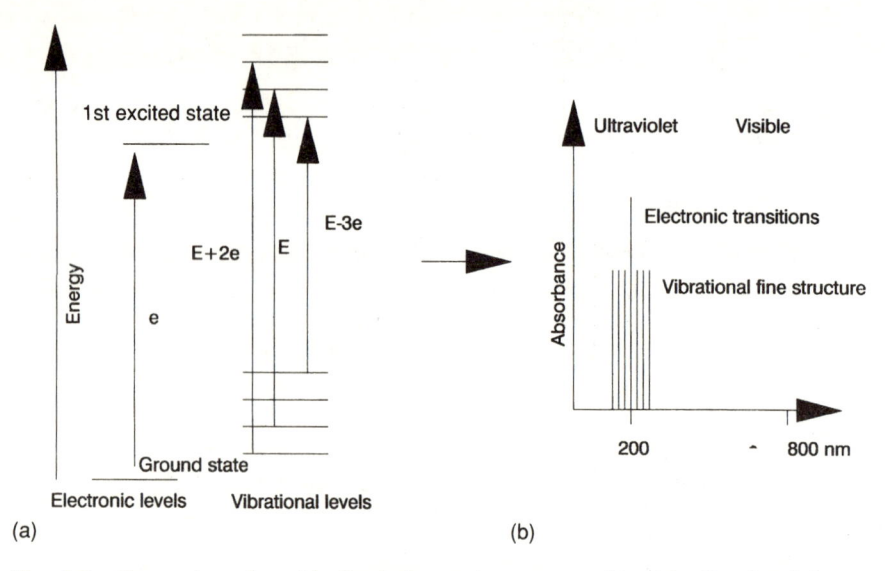

Fig. 1.6 Generation of an idealized electronic spectrum (b) with vibrational fine structure, from the molecular energy levels (a).

## 1.5 Laws of absorption of light

The conditions under which samples are examined are, by use of a blank, designed to minimize effects due to reflection, scatter and absorption by the solvent. The transmission of the solution and hence the absorption by the solute can thus be determined (Fig. 1.8).

The fractional amount of light absorbed and hence the fractional amount of light left for transmission are related to the thickness (Lambert, 1768) and concentration of the sample (Beer, 1852) [2]. Lambert's law may be derived as follows.

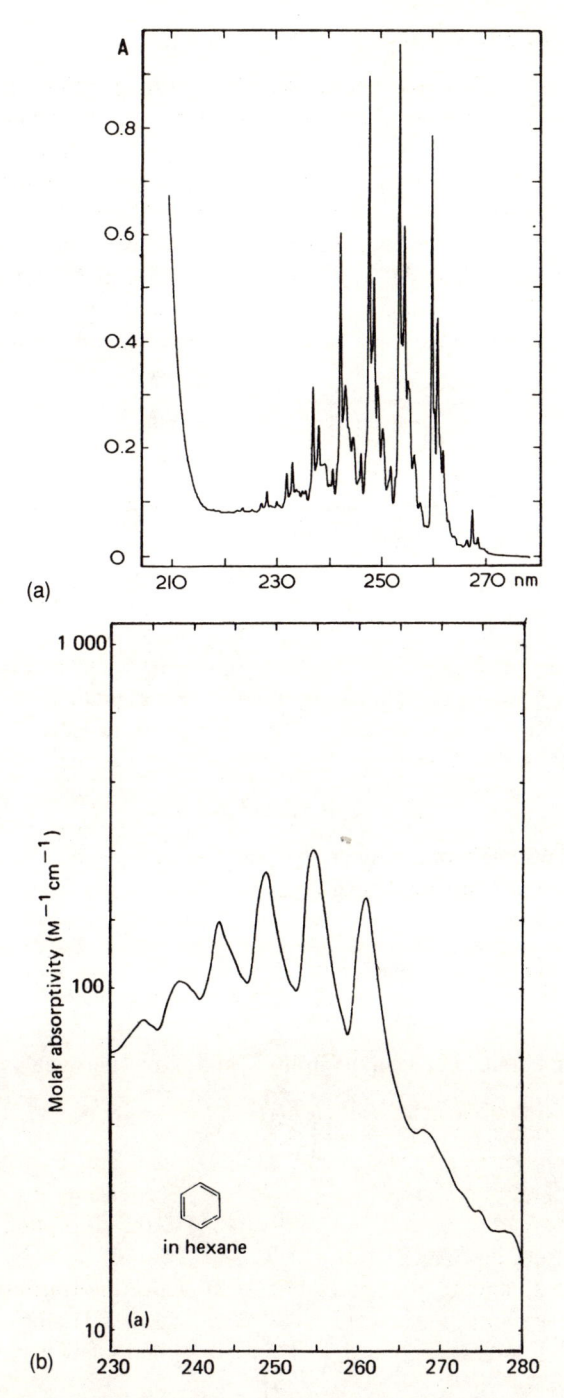

Fig. 1.7  Vapour phase (a) and solution (b) spectra of benzene.

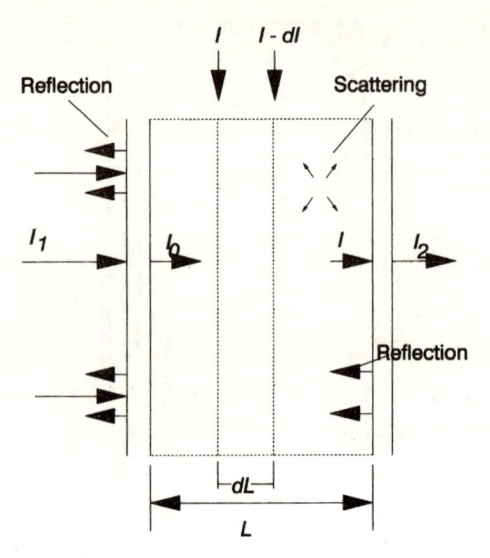

Fig. 1.8 *Transmission, reflection and scattering of radiation incident upon a cell of pathlength L.*

Let $I$ be the intensity of a parallel beam of monochromatic light of wavelength $\lambda$, passing through a layer of thickness $dl$ of an absorbing material. The change in intensity $dI$, is given by

$$dI = -k_\lambda I \, dl \qquad (4)$$

where $k_\lambda$ is a wavelength dependent proportionality constant. The negative sign is required because $I$ becomes smaller as $l$ becomes larger. The equation can be rearranged to give,

$$\frac{dI}{I} = -k_\lambda dl \qquad (5)$$

This can be integrated between the limits $I_0$ and $I$, for $l$ between 0 and $l$:

$$\int_{I_0}^{I} \frac{dI}{I} = -k_\lambda \int_{0}^{l} dl$$

$$[\ln I]_{I_0}^{I} = -k_\lambda [I]_{0}^{l}$$

$$\ln \frac{I}{I_0} = -k_\lambda l$$

$$2.303 \log_{10} \frac{I}{I_0} = -k_\lambda l$$

$$\log_{10} \frac{I_0}{I} = \frac{k_\lambda l}{2.303} \qquad (6)$$

If the concentration $C$ is the independent variable, as in Beer's law, we can write

$$dI = -k_\lambda I \, dC \tag{7}$$

Integrating between the limits $I_0$ and $I$, and $C = 0$ and $C = C$ gives

$$\log_{10}\frac{I_0}{I} = -\frac{k'_\lambda C}{2.303} \tag{8}$$

Equations 6 and 8 for $\log_{10} I_0/I$ in $l$ and in $C$ can be combined to give,

$$\log_{10}\frac{I_0}{I} = -\frac{k''_\lambda lC}{2.303} \tag{9}$$

$\log_{10} I_0/I$ is called *absorbance* ($A$), $I/I_0$ is called *transmittance* ($T$) and $k/2.303 = \varepsilon$ is called *molar absorptivity*, when the concentration ($C$) is given in g mol dm$^{-3}$. Thus,

$$A = \varepsilon lC \tag{10}$$

If the molecular weight of the sample material in solution is not known, the e*xtinction value* is used. This is defined as the absorbance, $A^{1\%}_{1\,cm}$ of a 1% w/v solution measured in a 1 cm cell. That is,

$$A^{1\%}_{1\,cm} = \frac{10\varepsilon}{M} \tag{11}$$

**Table 1.2** Beer–Lambert law and nomenclature*

| Accepted symbol | Meaning | Accepted name | Synonym+ | |
|---|---|---|---|---|
| | | | Abbreviation or symbol | Name |
| $T$ | $\dfrac{I}{I_0}$ | Transmittance | | Transmission |
| $A$ | $\log\dfrac{I_0}{I}$ | Absorbance | $OD, D, E$ | Optical density, extinction |
| $A^{1\%}_{1cm}$ | $\dfrac{10\varepsilon}{M}$ | Extinction value | $E^{1\%}_{1cm}$ | |
| $a$ | $\dfrac{A}{bc}$ | Absorptivity | $k$ | Extinction coefficient, Absorbancy index |
| $\varepsilon$ | $\dfrac{A}{bM}$ $M = $ mol dm$^{-3}$ | Molar absorptivity | $a^M$ | Molar extinction coefficient |
| $b$ | Path length of radiation through sample | Cell path | $l$ or $d$ | |

* Based on recommendations in *Anal.Chem.* **46** (1973) 2449
+ Not recommended but occasionally used in the literature

10  *Principles of spectrophotometric measurements*

Various alternative terms of absorbance and symbols appear in the literature (Table 1.2). There is at present no IUPAC convention for symbols for use in molecular absorbance spectroscopy.

## 1.6 Quality of a spectrum

The variation of molar absorptivity, $\varepsilon$, with wavelength, $\lambda$, and hence absorbance, A, with wavelength, at a fixed concentration, $c$, and path length, $l$, is called a spectrum.

The quality of a spectrum obtained using a dispersive instrument is a function of spectral purity or monochromaticity of the probe radiation. This depends on the physical width of the slits used to isolate the probe radiation from the source's spectrum (Fig. 1.9) and of the amount of stray light present. It is convenient to discuss the effect of physical slit width and of stray light separately.

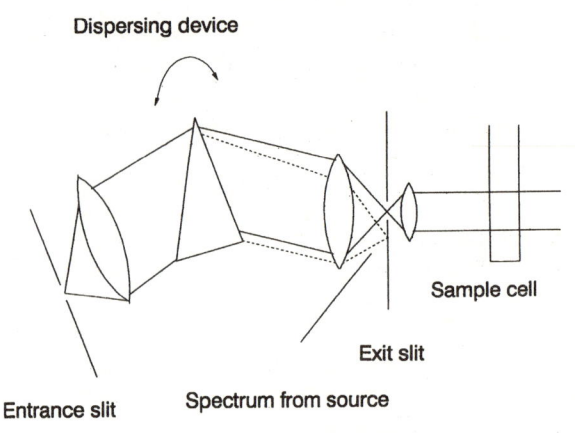

Fig. 1.9 *Production of a beam of radiation whose profile is determined by the physical width of the slits.*

### 1.6.1 *Effect of slit width on spectra*

As the width of the slits is narrowed, the shape of the bands within a spectrum alters until a more-or-less constant picture emerges, as illustrated for the spectrum of benzene (Fig. 1.10). However, at very narrow slit widths the amount of energy reaching the photomultiplier detector is reduced and noise may disturb the spectral quality of the spectrum as shown for cytochrome C (Fig. 1.11).

The physical width of the slits determines the *spectral band width* of the probe radiation emergent from the monochromator. Consider a monochromator whose entrance and exit slits are 1 mm wide and whose optical design permits the exit slit to be filled by the entrance slit image. Assume that a monochromatic source of wavelength $\lambda_0$ is incident on the monochromator

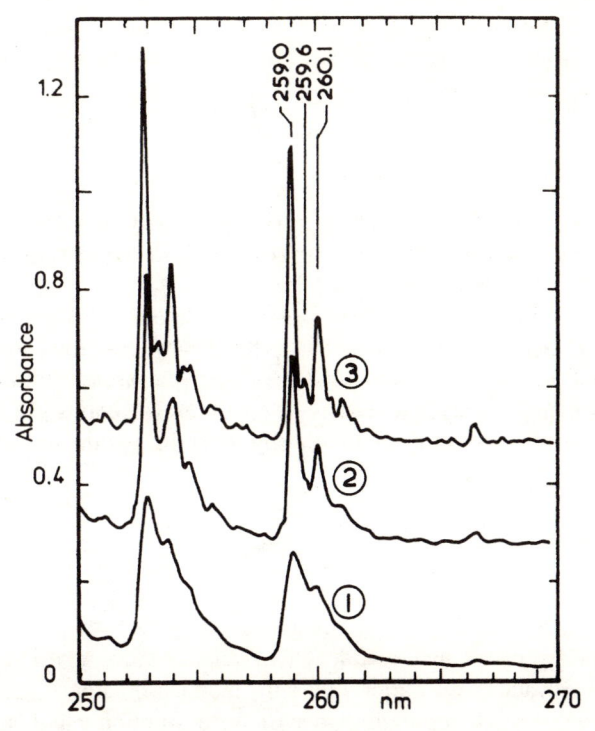

Fig. 1.10  *Effect of spectral slit width on the spectrum of benzene in cyclohexane. (1) SSW = 2.0nm, (2) SSW = 1.0nm, (3) SSW = 0.5nm. Curves displaced vertically for clarity.*

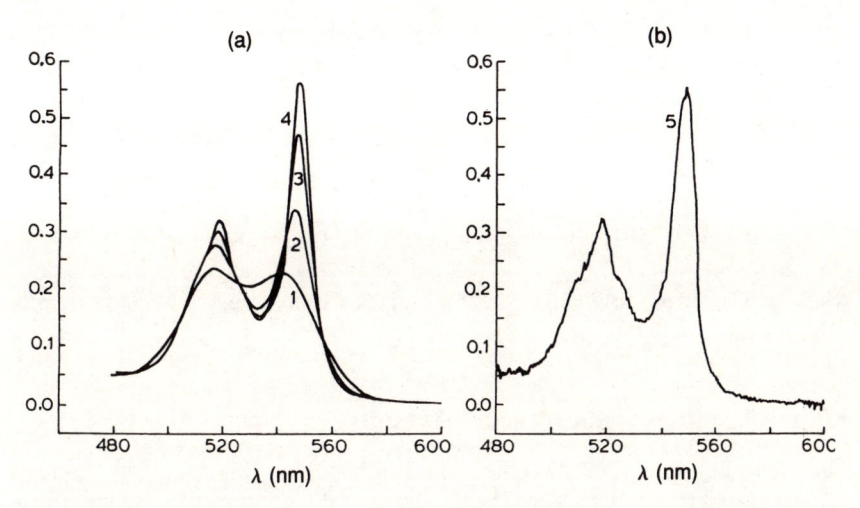

Fig. 1.11  *Effect of spectral slit on the spectrum of cytochrome C. Bandwidths: (1) 20, (2) 10, (3) 5, (4) 1, and (5) 0.08nm. Note that the reduction in band width from 1 to 0.08nm shows an increase in noise level but with no noticeable change in peak positions or shapes.*

## 12  Principles of spectrophotometric measurements

whose linear reciprocal dispersion is 20 nm per mm (i.e. the wavelength of dispersed light varies by 20 nm for each 1 mm travel along the spectrum). When the dispersing element (prism or grating) is rotated, the image of the entrance slit, hatched on Fig. 1.12(a), can be made to move along the focal plane and across the exit slit. The progress of this image, shown on the diagram, generates a triangular relative intensity variation with wavelength as shown in Fig. 1.12(b). The width at half the intensity peak height is called the *effective spectral bandwidth*. It can be measured by scanning a sharp line source such as that from a mercury lamp.

When a (white light) source replaces the monochromatic source, the monochromator extracts an equivalent shaped band centred at the monochromator's setting, with the wavelength band indicated by the spectral bandwidth. Provided the spectral band width is narrow as compared to the detail within the spectrum, the spectrum is not very dependent on bandwidth and the recorded spectrum approaches the true spectrum.

### 1.6.2  Effect of stray radiation on spectra

Spectra may also be distorted by the monochromator's emergent beam containing light quite different to that indicated by the setting of the wavelength scale, allowing for the band width. This is the so-called *stray light*, caused by scatter and unwanted reflections within the monochromator. If the instrument is set at a wavelength corresponding to a known absorption maximum ($\lambda_{max}$) for a sample, the stray light will, in general, be less absorbed by the sample than that at other wavelengths and thus the amount of light absorbed at $\lambda_{max}$ will be too low.

If $S$ is the intensity of stray radiation unabsorbed at a wavelength $\lambda_0$, and $I_0$ and $I$ are the incident and transmitted intensities of light at $\lambda_0$, the apparent absorbance $A_{app}$ is given by

$$A_{app} = \log_{10}\left(\frac{I_0 + S}{I + S}\right) \quad (12)$$

Thus, if $S = 1\%$ of $I_0$ the maximum value of $A_{app} = 2$ and for $S = 0.1\%$ of $I_0$ the maximum value of $A_{app} = 3$. The net result over a range of concentrations where $I$ varies from low values up to $I_0$ gives rise to a curved Beer's law plot (Fig. 1.13).

### 1.7  Accuracy and precision of results

The closeness of a spectroscopic value (either $A$ or $\lambda$) to the true value is expressed in terms of accuracy. The precision of the set of results used to obtain the mean value (best estimate to the true result from the experiments carried

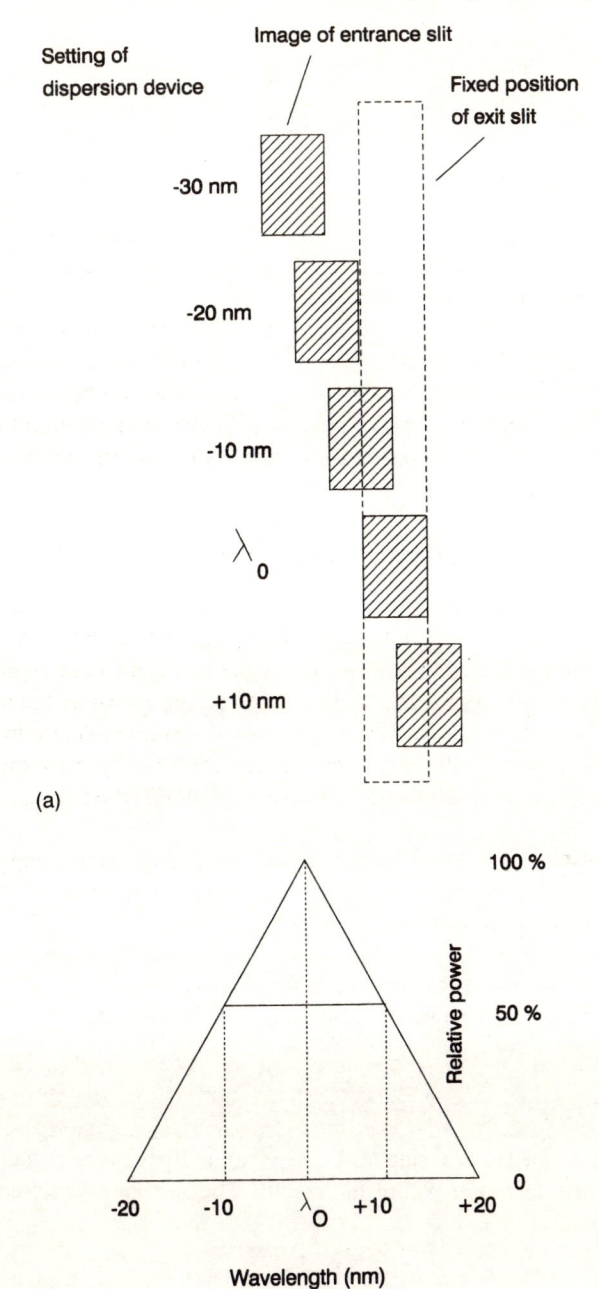

Fig. 1.12 *Origin of the triangular emergent beam energy distribution with wavelength (b) from scanning monochromatic radiation of wavelength $\lambda_0$.*

## 14 Principles of spectrophotometric measurements

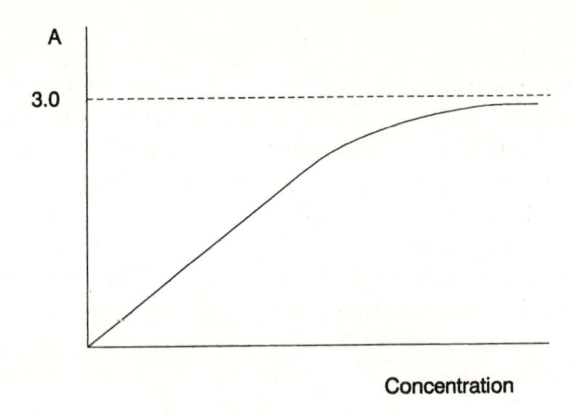

Fig. 1.13 *Effect of 0.1% stray radiation on a Beer's law plot.*

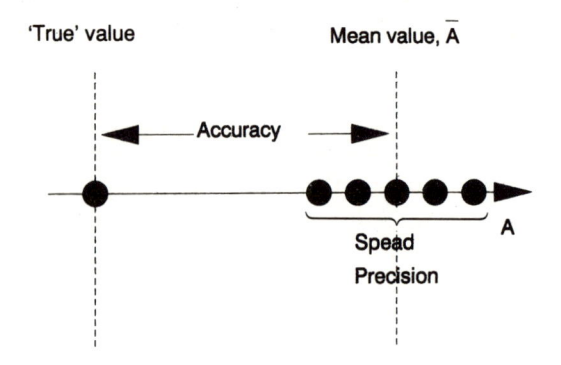

Fig. 1.14 *Concepts of accuracy and precision.*

out) is indicated by the spread of the results (Fig. 1.14).

The mean value, $A$, for a set of results $A_1 \ldots A_n$ is given by

$$A = \frac{A_1 + A_2 + A_3 + \ldots A_n}{n} \quad (13)$$

The precision of the set of results $A_1 \ldots A_n$ is normally expressed mathematically by the *standard deviation*, $S_A$, of the results. The standard deviation is defined by the equation

$$S_A = \sqrt{\frac{\Sigma(A_i - A)^2}{n-1}} \quad (14)$$

where $A_i$ is the *i*th value of the data set and *i* varies from 1 to *n* along the data set.

## 1.8 Production of a valid spectrum

In order to obtain a valid (near true as possible) spectrum, it is necessary to measure absorbance and record wavelength values with sufficient accuracy and precision. The precision of results is a function of the general experimental care taken and of the number of replicates (being proportional to $n^{-1/2}$) taken. The accuracy depends on the following instrumental factors: the accuracy of the absorbance and wavelength scales; a low stray light figure; and the spectral band pass being narrow compared to the fine detail in the spectrum to be measured.

If it is intended to evaluate the molar extinction coefficient, it is necessary to have accurate and precise values for the path length, $l$, and concentration of the absorbing species, $C$, since,

$$\varepsilon = \frac{A}{Cl} \tag{15}$$

The problems associated with determination of path length and concentration are discussed later (Chapter 2).

## 1.9 Standards

Standards are used to ensure that the performance of an instrument with respect to wavelength, absorbance scale and stray light characteristics is acceptable. These 'links' with 'sanity' (reality) [3] are discussed later in Chapter 3 and, in more detail, in C. Burgess and A. Knowles [4].

## 1.10 Summary

This chapter has discussed the nature of radiant energy and its interaction with matter. Its frequency, wavelength and wavenumber, units, the origin and nature of spectra, and the laws of the absorption of light are presented. The effects of slit width and stray light on the quality of spectra are also outlined. Some useful texts are listed below.

## References

1  Boyle, R. (1664) *Experiments and Considerations Touching Colours ...*, H. Herringman, London, p. 91.
2  Thorburn Burns, D. (1987) Aspects of the development of colorimetric analysis and quantitative molecular spectroscopy in the ultraviolet–visible region, in *Advances in Standards and Methodology in Spectrophotometry*, (eds C. Burgess and K.D. Mielenz), Elsevier, Amsterdam, p. 10.
3  Edisbury, J.R. (1966) *Practical Hints on Absorption Spectrophotometry 200–800 mμ*, Hilger and Watts, p. 172.
4  Burgess, C., and Knowles, A. (eds.) (1981) *Standards in Absorption Spectrometry*, Chapman and Hall, London, Chapters 4–7.

## Further reading

*General, theory*

Ingle, J.D. and Crouch, S.R. (1988) *Spectrochemical Analysis*, Prentice Hall, Englewood Cliffs, NJ. (Good basic text on all areas of spectroscopy.)

Olsen, E.G. (1975) *Modern Optical Methods of Analysis*, McGraw Hill, New York. (A readable general text.)

*UV–visible, theory*

Nowicka-Janowska, T., Gorczynaska, K., Michalik, A., and Wieteska, E. (1986) *Analytical Visible and Ultraviolet Spectrometry*, Elsevier, Amsterdam. (Scholarly theory, instrumental and practical aspects.)

Schenk, G.H. (1973) *Absorption of Light and Ultraviolet Radiation: Fluorescence and Phosphorescence Emission*, Allyn and Bacon. (Readable).

Meehan, E.J. et al. (1981) *Treatise on Analytical Chemistry*, Part I, Vol. 7, 'Optical Methods of Analysis', Interscience, New York. (Detailed theory including equipment.)

*Instrumental and practical aspects*

Lothian, G.F. (1969) *Absorption Spectrophotometry*, 3rd edn, Hilger, Bristol. (Still a useful text.)

Strobel, A.A. and Heineman, W.R. (1989) *Chemical Instrumentation: A Systematic Approach*, 3rd edn, Wiley, New York. (Excellent account of instrumental aspects.)

Miller, J.N. (ed.) (1981) *Standards in Fluorescence Spectrometry*, Chapman and Hall, London. (Essential reading.)

Knowles, A., and Burgess, C. (ed) (1984) *Practial Absorption Spectrometry*, Chapman and Hall, London. (Essential reading.)

Burgess, C., and Mielenz, K.D., *Advances in Standards and Methodology in Spectrometry*, Elsevier, Amsterdam. (Modern aspects apart from historical introduction.)

# 2 Instrumentation for UV–visible and fluorescence spectroscopy

## 2.1 Introduction

This chapter covers the laboratory aspects of obtaining absorption and fluorescence spectra and of making quantitative measurements in these areas. There is some emphasis on fluorescence measurements, since both in theory and in practice fluorescence results are more dependent upon the experimental conditions than absorption spectrometry. Also, there are added complications which include the several different types of fluorescence spectra available and the correction of these spectra. The emphasis throughout is on optical aspects, with data collection and data handling perspectives covered in later chapters.

## 2.2 Instrumentation for absorption spectrometry

In UV–visible absorption spectrometry, light from a suitable source is passed through a prism or grating monochromator (filter instruments are still often used), then passed through the sample before reaching the detector (Fig. 2.1a). This arrangement, which is the opposite of that used in infrared spectrometry, minimizes exposure of the sample to light of wavelengths that might cause photodecomposition. In contrast, however, 'reverse optics' are used in linear photodiode array instruments (Fig. 2.1b). In the simplest instruments, these components are in a single beam sequence and such devices are cheap, robust and very simple to maintain and use. Since the light source, output and detector response will be wavelength dependent, such instruments are only used for single wavelength studies where a reference sample is measured separately from the test sample. Spectral 'scans' can now be achieved with the aid of specialized software (and with the assumption of temporal stability in the instrument components).

Double beam spectrometers facilitate the measurement of absorption spectra, as the light from the source is split into two equal beams after passing through the monochromator (Fig. 2.1c). The two beams pass through the test sample and reference sample (their light paths being otherwise the same) and the spectrum obtained is thus compensated for changes in some of the system variables (e.g. source intensity). In most modern instruments the two beams are recombined

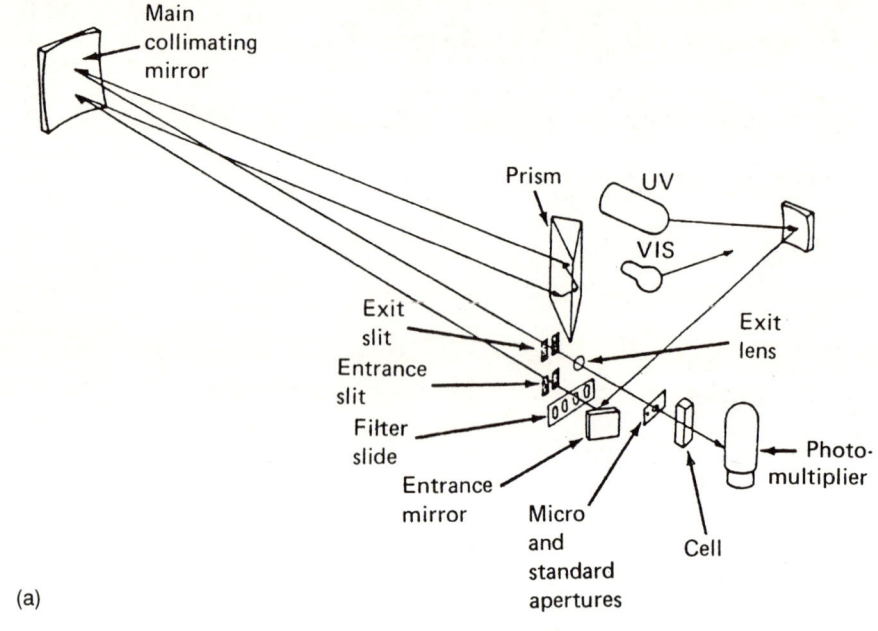

Fig. 2.1 *Optical layout of typical absorption spectrometers. (a) Single beam (by courtesy of the Shimadzu Corporation).*

before detection by a single detector ('double beam in time'). The beams can be distinguished for example by an LED and auxiliary detector which monitors the beam-splitter position. Two separate detectors ('double beam in space') can also be used, but this is more complex and less common in practice.

## 2.3   Instrument components in absorption spectrometry

The crucial optical components in UV–visible spectrometry are the light source, the monochromator, the beam splitting system (if applicable) and the detector. In addition all instruments include mirrors and some utilize lenses and therefore the properties of these components are also important.

In contrast to fluorescence spectrometry, an intense light source is not a prerequisite in absorptiometry: stability and coverage of the spectral range are more important. Normal light sources are the deuterium arc and the tungsten filament lamps for lower and higher wavelength regions, respectively. A 50 W deuterium lamp, typically operating at about 100 V, 500 mA, gives a usable output from about 190 to 350 nm. At higher wavelengths its intensity is rather feeble. Conveniently, a tungsten lamp (typically up to 100 W) whose output approximates to that of a blackbody emitter, becomes useful at about 330 nm and higher wavelengths. The so-called quartz–halogen lamps have higher operating temperatures for elevated output and quartz envelopes for better UV

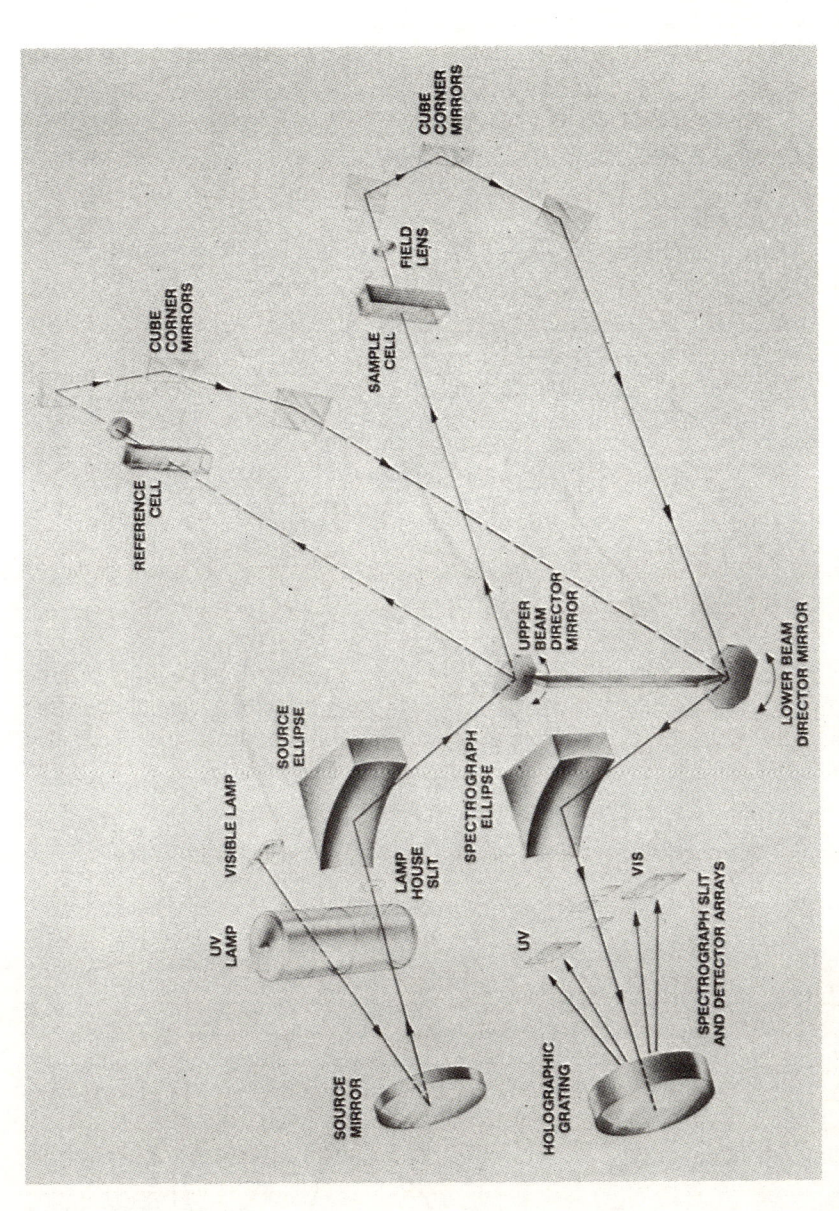

Fig. 2.1 Optical layout of typical absorption spectrometers. b) Diode array reverse optics (by courtesy of Hewlett-Packard Ltd).

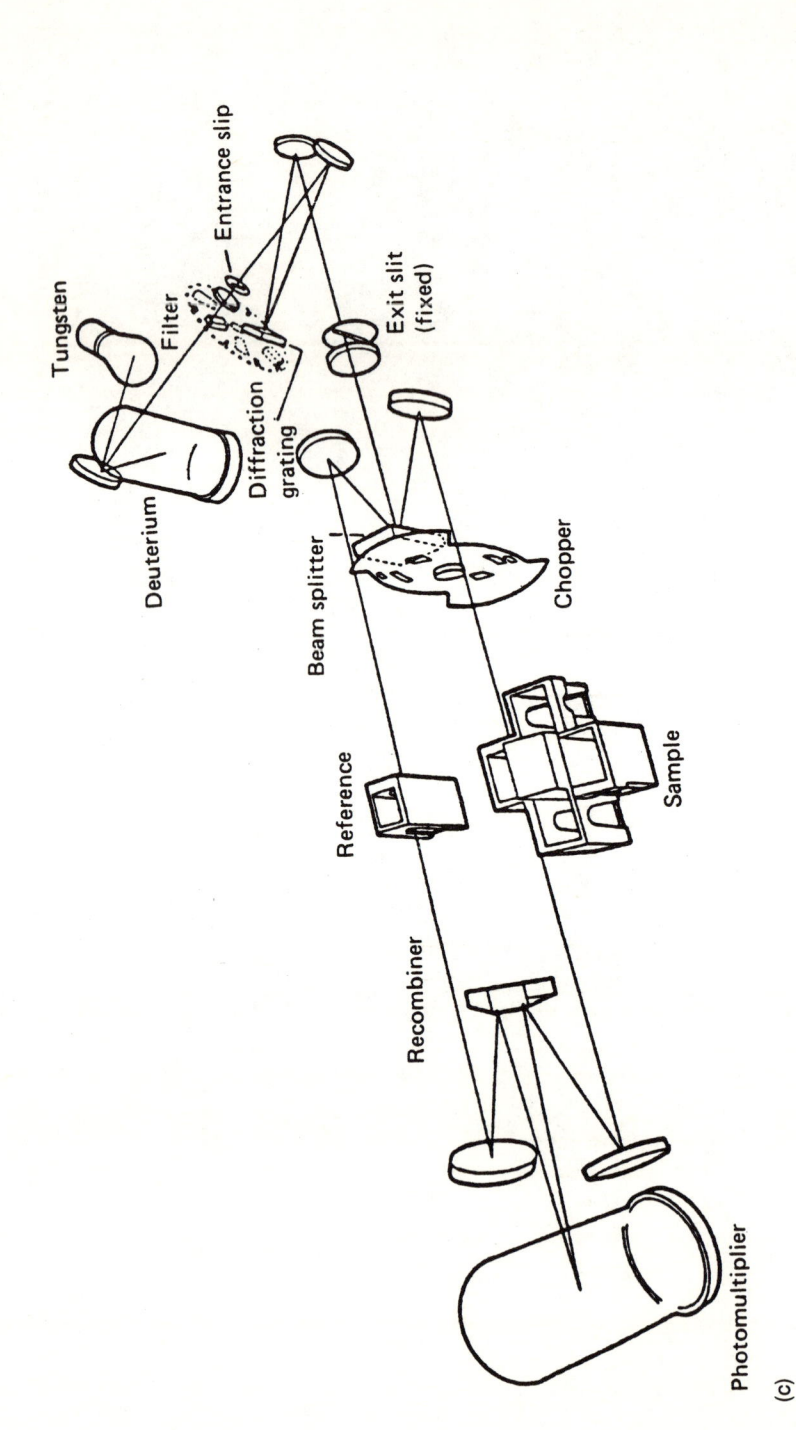

Fig. 2.1 Optical layout of typical absorption spectrometers. (c) double beam (by courtesy of Bausch and Lomb).

transmission. The presence of the halogen, which combines with evaporated tungsten, prolongs the lifetime of the lamp by reducing blackening of the envelope (Fig. 2.2).

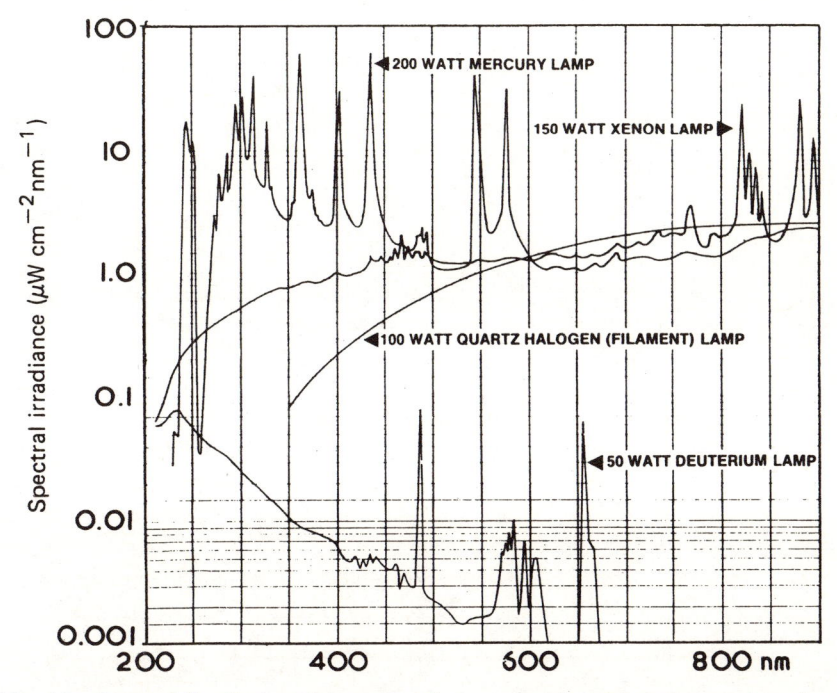

Fig. 2.2 *Spectral distribution of the outputs from commonly used arc lamps: deuterium, xenon and mercury. Reproduced by courtesy of the Oriel Corporation.*

Most modern instruments use grating monochromators in reflection mode as dispersing elements. The design of the monochromator produces compact optical paths and reduces optical aberrations. The Ebert, Czerny–Turner and Littrow mountings (Fig. 2.3) are amongst the most common. Double monochromators are sometimes used, and they are recommended when high absorbance values are measured and stray light must be minimized. They involve, however, a considerable reduction in light throughput, and must, therefore, result in a degradation of the $S:N$ ratio. Most gratings in UV–visible spectrometers are blazed (i.e. the grooves cut at an angle) to give maximum efficiency (usually 60–70%) at about 250 nm. Holographic (i.e. optically ruled gratings) are now commonly found and give significantly better stray light characteristics than mechanically produced gratings. Concave holographic gratings have been used in some designs with the aim of reducing the number of optical components. Auxiliary filters may be included in instruments designed for use at high wavelengths, as it is necessary to remove higher order signals transmitted at 1/2, 1/3 etc of the first-order wavelength.

(a)

Spherical mirror

(b)

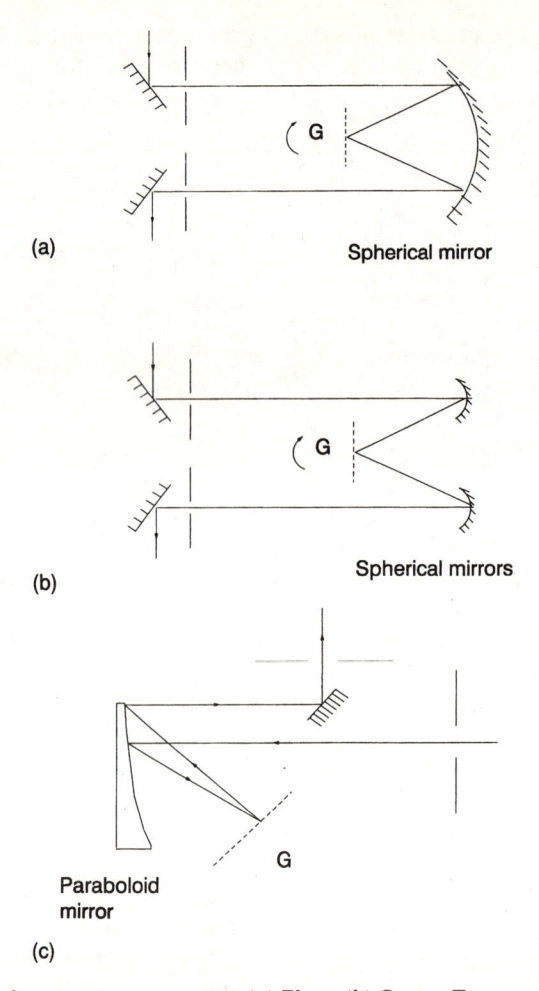

Spherical mirrors

(c)

Paraboloid
mirror

Fig. 2.3  *Monochromator arrangements. (a) Ebert, (b) Czerny-Turner, and (c) Littrow.*

Several types of beam splitter have been used in commercial double-beam spectrometers. These include rotating sector mirrors and half-silvered mirrors with, in almost all cases, provision made for a 'dark' period (i.e. the detector receives no light from the source via either beam). This is designed so that the dark current of the detector can be measured. The properties of the reflecting surfaces in any spectrometer are crucial; the light beam might strike 10 or more such surfaces during its passage from source to detector and the overall attenuation is large even if only 3–5% of the light is lost at each reflection. In practice, however, reflectivities of >90% are hard to achieve, especially at around 200 nm (aluminium front-surface mirrors) and it is also in this region that the properties of the surfaces deteriorate with time most rapidly.

Sample cells are available in a wide variety of types and materials with most routine measurements using 10 mm open-top rectangular cells. They may be fabricated from acrylic plastic, in which case they are inexpensive and disposable, although they are not always solvent resistant and not suitable at wavelengths below about 300 nm. Glass cells are better, but not below about 320 nm. The best are cells of fused quartz, which are suitable to about 210 nm, and fused synthetic silica cells are suitable above 190 nm. In practice only a small portion of a 10 mm cell is actually 'used' in the optical measurement system of most spectrometers and therefore smaller cells may be used to economize on sample. Numerous designs and sizes of flow cell are also available.

For most purposes, cells in absorptiometry are not thermostatted, although temperature changes will affect molar absorptivities (a common but slight effect only), molecular equilibria, solvent density and evaporation. The most usual reason for controlling the temperature is in the study of enzyme-catalysed reactions, where temperature effects may be substantial.

The condition of the cells is also an important factor and the cells can be cleaned with nitric acid, although in absorptiometry detergents and organic solvents may prove suitable and milder alternatives.

Detectors in routine spectrometry are commonly, yet wrongly, named 'photomultipliers' (they actually multiply electrons!). Although sensitive and robust for long-term use, they are quite costly and require a stable high voltage power supply. Most of the commonly used photocathodes have maximum sensitivity at about 400 nm with a sharp fall-off above 600 nm. However, 'red-sensitive' tubes are readily available and these extend the useful range up to about 900 nm.

In single beam spectrometers, phototubes which are non-amplified devices with receptors similar to photomultipliers may be satisfactory. Solid state detectors (e.g. silicon photodiodes) are also used in simple instruments. Multichannel detectors are treated in Chapter 8.

## 2.4 Instrumentation for luminescence spectroscopy

Luminescence may be defined as light, other than blackbody radiation, emitted by a sample. Before it can occur, an initial excitation step must take place. In this section emphasis is placed on *photoluminescence*, i.e. luminescence in which the excitation is achieved using an external light source. Several photoluminescence phenomena have been identified, although in practice, only fluorescence and to a lesser extent phosphorescence find practical use.

Methods based on luminescence phenomena have a number of important advantages.

1. *Versatility.* A very large range of samples can be studied — organic and inorganic, synthetic and naturally occurring, small and large molecules. Samples may be in dilute or concentrated solutions, gases, suspensions, or solid surfaces. Applications may be simply analytical, or may involve studies of molecular structure and interactions, or the location of a species. Luminescence methods can often be combined very advantageously with other methods, for example High Performance Liquid Chromatography (HPLC), Thin Layer Chromatography (TLC) and microscopy.
2. *Sensitivity.* Most users of fluorimetry and phosphorimetry are attracted by the high sensitivity of the methods. In solution studies, pg ml$^{-1}$ levels can often be determined, which is in contrast to the $\mu$g ml$^{-1}$ levels that can be detected in absorption spectroscopy. Instrumental advances are continuing, and so may be expected to improve sensitivity still further.
3. *Selectivity.* Luminescence techniques are generally more selective than absorption methods, since two distinct wavelengths (those of absorption and emission) can be used to characterize a sample. If phosphorimetry is used, then other parameters, including the lifetime of phosphorescence and the *P:F* ratio, are available. In studies of complex mixtures such as those occurring in biochemistry, further selectivity may be required and can be provided by using spectroscopic deconvolution methods (e.g. derivative spectroscopy), by association with separation methods, or by combination with specific biochemical reactions (e.g. immunoassays, enzyme assays).

## 2.5 Origin of molecular luminescence

When UV or visible light is used to irradiate a sample, a number of phenomena may occur (Fig. 2.4). In many cases most of the photons will pass straight

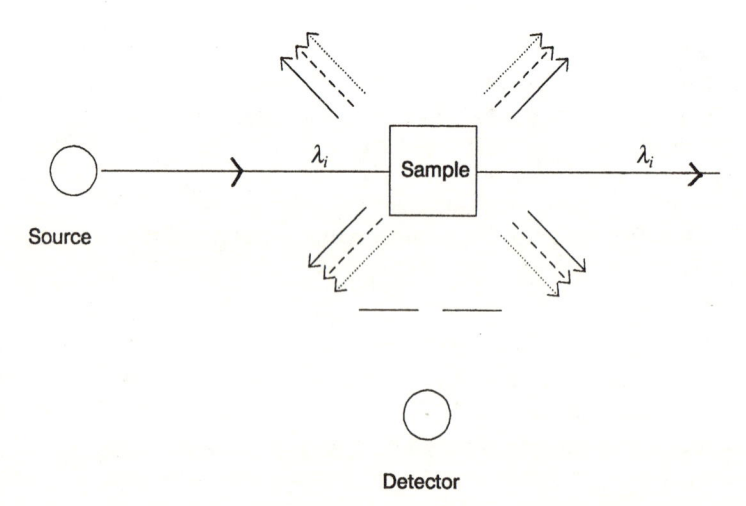

Fig. 2.4 UV–visible optical phenomena. Photon energy, $E = hc/\lambda$ (———▶) Rayleigh scattering, $\lambda_{sc} = \lambda_i$ (---▶); Raman scattering, $\lambda_{sc} > \lambda_i$ or $< \lambda_i$ (········▶); fluorescence, $\lambda_f > \lambda_i$.

*Origin of molecular luminescence* 25

through the sample, in contrast to absorption spectroscopy where the observation of such photons is not required. The luminescence phenomena are normally observed at right angles to the incident beam. Some of the photons are scattered, i.e. deviated from their original courses by collisions with other bodies. The two scattering phenomena of most concern are (i) Rayleigh scattering, in which the wavelength of the light remains unchanged (i.e. elastic collisions), and (ii) Raman scattering, which involves a wavelength shift, and which may thus be confused with fluorescence. Raman scattering is normally a feeble effect compared with Rayleigh scattering. A small proportion of the photons may be absorbed by the sample, causing the molecule to gain both electronic and vibrational energy (Fig. 2.5).

An excited molecule will lose its energy rapidly by one or more of a number of pathways. The pathway that predominates will be that with the most favourable rate constant. Within about $10^{-12}$ s, the processes of internal

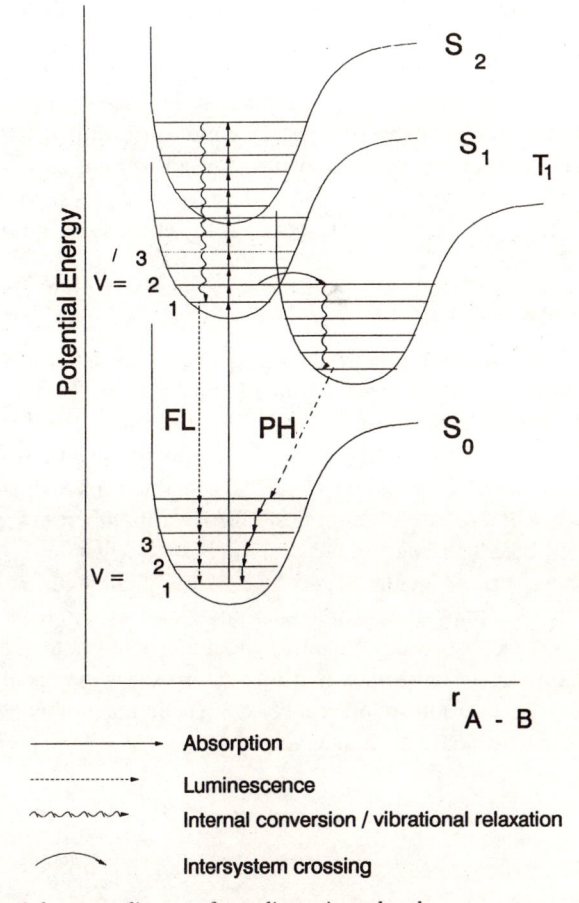

Fig. 2.5 *Potential energy diagram for a diatomic molecule.*

conversion and vibrational relaxation will bring the molecule to the lowest vibrational level ($V^1 = 0$) in the first excited singlet electronic level, $S_1$. Here the molecule may remain for about 10 ns before returning to the ground state via a radiative transition called *fluorescence*. The emission will evidently have a longer wavelength, i.e. a lower energy, than the absorption transition. This wavelength difference, to which solvent relaxation effects may also contribute (see below), is sometimes called the Stokes shift. A molecule in the $S_1$ state may also undergo intersystem crossing to the lowest triplet state, $T_1$, which has a lower energy than $S_1$. A subsequent radiative transition from $T_1$ to $S_0$ is the basis for phosphorescence. Because S–T transitions are forbidden, the lifetime of phosphorescence is very long ($10^{-3}$–$10^2$ s). This phenomenon is generally observed at 77 K: at room temperature it will only be observable in practice when the sample is absorbed on a solid surface, or when the molecule is protected from collisional quenching, e.g. by being enclosed in a micelle. Other mechanisms by which a molecule may lose its excitation energy include photodecomposition and nonradiative energy transfer. Energy transfer is also common, especially in biological systems and in molecules containing more than one chromophoric group.

The types of molecule most likely to show useful fluorescence are those with delocalized $\pi$-orbital systems. Substituents that increase the electron density in such systems normally enhance fluorescence intensity while electron-withdrawing groups reduce fluorescence intensity. Certain groups, for example Br, I, exert a 'heavy-atom' effect and enhance the rate of intersystem crossing. Such groups may thus enhance phosphorescence at the expense of fluorescence.

## 2.6   Fluorescence intensity

The excellent sensitivity of fluorimetry and related methods is due to the fact that the fluorescence signal from a dilute solution is detected against a 'dark' background. In the absorptiometry of dilute solutions, a small difference between two substantial light signals is sought; this is much more difficult. With the aid of a number of gross assumptions, we can determine the factors affecting the fluorescence of a dilute solution. From the Beer–Lambert law, the amount of light absorbed by a solution is given by $I_0 (1-10^{-\varepsilon lc})$, where $I_0$ is the incident light intensity, $\varepsilon$ the molecular absorptivity, $l$ the pathlength, and $c$ the molar concentration. We may define the quantum yield, $\varphi_f$, of the sample as the number of photons absorbed, divided by the number of photons emitted, i.e. as the fraction of excited molecules that lose their excess energy by fluorescing. (The quantum yield of phosphorescence, $\varphi_p$, has an analogous definition). The intensity of fluorescence, $I_f$, is thus given by

$$I_f = I_0\, \varphi_f(1-10^{-\varepsilon lc})$$
$$= I_0\, \varphi_f \varepsilon lc$$

if $\varepsilon lc$, the absorbance, is small.

In principle, therefore, the observed fluorescence intensity, which will only be a fraction of $I_f$ because of the instrumental factors, is proportional both to the solute concentration and to the intensity of the exciting light source at the absorption wavelength. In practice this equation only holds in a very limited set of circumstances; for example, if the absorbance of the solution is only 0.05, a deviation of 5% from linearity can be expected. Further deviations from a linear analytical growth curve arise because of the inner filter effect (Fig. 2.6). The intensity of the observed fluorescence is reduced by absorption of energy both before and after the small central portion of the sample cell in use. Absolute determinations of the quantum yield, $\varphi_f$, are difficult to perform and it

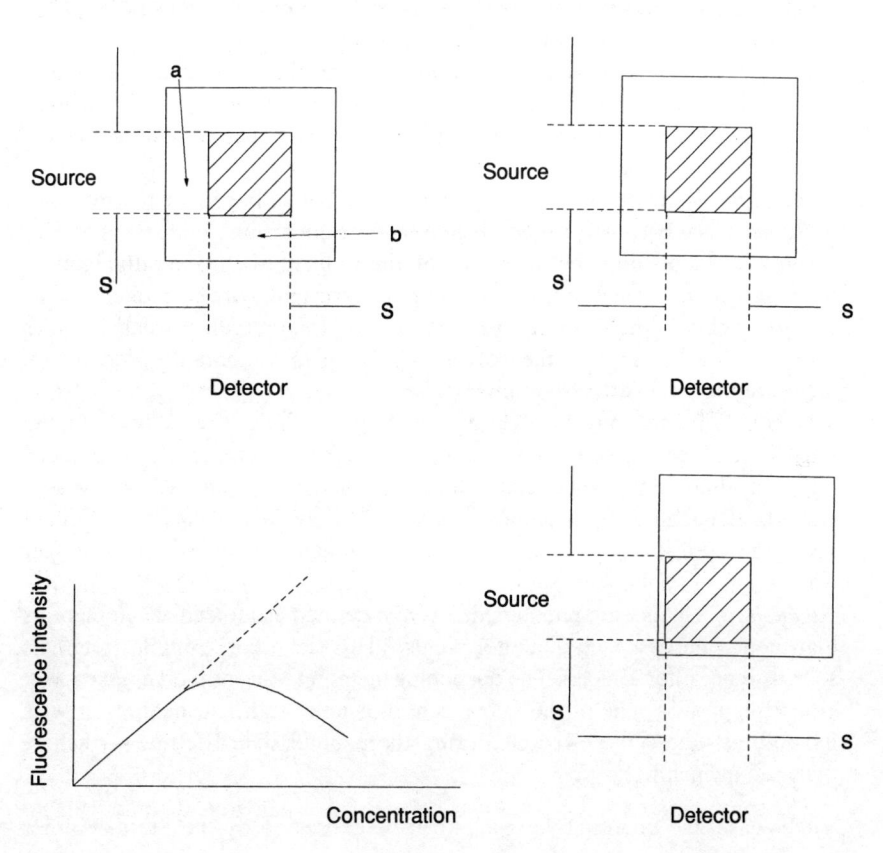

Fig. 2.6  *Inner filter effects. Key: (a) prefilter effect; (b) postfilter effect; (s) slits. Also, fluorescence intensity plotted against concentration showing deviation from linear growth.*

is essential to use corrected spectra, to ensure the absence of inner filter and energy transfer effects and to take into account fluorescence polarization phenomena. However, the *relative* quantum yields of a series of chemically-related molecules may be easier to determine, and of some value in studying structure-luminescence correlations.

## 2.7 Factors affecting fluorescence intensity

There are several factors that affect fluorescence intensity. These are:

1. *pH*. A large number of fluorescent species contain ionizable groups. In such cases it is common to find that only one ionic form of the molecule is fluorescent, and thus that pH control is important. For example, studies of the oxybarbiturates have shown that only the dianions are fluorescent (pH 13). It is also noteworthy that many protonation reactions have rate constants higher than that of fluorescence. It is thus possible to observe the absorption spectrum of the neutral molecule and the fluorescence spectrum of the ionised molecule (e.g. β-naphthol at pH 3).
2. *Temperature*. Increasing the temperature will normally reduce the intensity of fluorescence because of increased collisional quenching.
3. *Viscosity*. Increasing the viscosity of the solvent will generally lead to increased fluorescence, since collisional interactions will be reduced. This procedure is only (deliberately) used in studies of fluorescence polarization as shown below. Note that the solvent viscosity is temperature dependent, therefore these two effects are interactive.
4. *Solvent*. Solvents may have large and unpredictable effects on both the intensity and the wavelength of fluorescence. Intensity effects may occur as a result of alterations in the relative energies of the electronic states, or as a result of quenching phenomena as discussed below. Wavelength effects may occur via the Lippert mechanism (Fig. 2.7) which predicts a shift to longer wavelengths in polar solvents.
5. *Quenching*. Quenching phenomena may be defined as molecular interactions that reduce fluorescence quantum yields. (This definition excludes artefacts such as inner filter effects). The quenching agent, Q, can form complexes with either the ground state of the fluorescent molecule (static quenching) or with its excited state (via collisions during the excited state lifetime — hence, dynamic quenching).

In either case, the quantitative results can be expressed by the Stern–Volmer equation:

$$\frac{F_0 - F}{F} = K[Q]$$

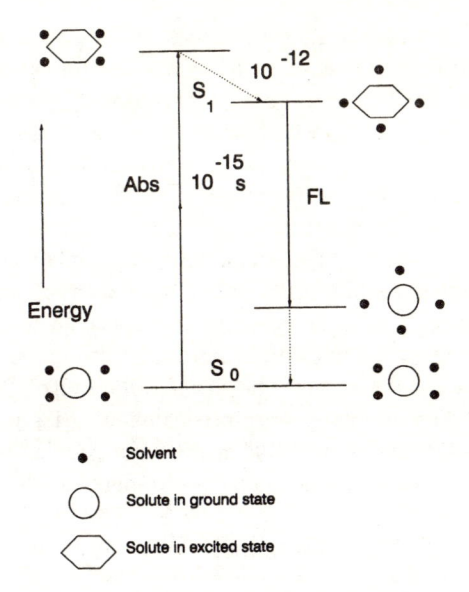

Fig. 2.7 *Solvent effects on fluorescence wavelength.* (———) *Radiative transitions;* (----) *non-radiative transitions.*

In the case of static quenching the constant $K$ is related to the equilibrium constant of the complex; in the case of dynamic quenching, $K$ is the product of the rate constant of the quenching process and the natural fluorescence lifetime.

## 2.8 Fluorescence polarization

It has been known for almost 60 years that fluorescence signals can be polarized, i.e. yield signals of different intensity when the sample is observed through crossed and through parallel polarizers. A small molecule in a non-viscous solvent normally shows entirely depolarized fluorescence, since the molecule suffers many collisions and random motions in the approximate 10 ns between the absorption and emission of light. A large molecule, or a small molecule that is immobilized by being bound to a large molecule or studied in a viscous medium, will show polarized fluorescence.

Polarization is normally measured by illuminating the sample with vertically polarized light and observing the emitted fluorescence sequentially through a second vertically oriented polarizer (giving a fluorescence intensity $I_{II}$) or a horizontally oriented polarizer (giving an intensity $I_I$). The polarization, $p$, is then given by

$$p = \frac{I_{II} - I_I}{I_{II} + I_I}$$

## 30  Instrumentation for UV–visible and fluorescence spectroscopy

Polarization attachments are available for many fluorescence spectrometers, and are easily fitted and used. It is wise to check their performance before use in the spectral range under study (for example, do they really give zero transmission when 'crossed'?). It should also be remembered that absolute values of $p$ are affected by the polarization properties of the spectrometer's emission monochromator. Corrections for such effects can be made if necessary, but often only changes in polarization are of interest.

In practice, polarization phenomena have a number of practical uses. Substances with otherwise similar fluorescence properties may have different degrees of polarization, perhaps because of size differences. Polarizers can be used to study the rotational relaxation times of macromolecules and to study molecular interactions. It is found that the interference of Rayleigh scattered light can often be preferentially removed using polarizing films. Circularly polarized fluorescence, the excited state analogue of circular dichroism, has also found use recently in studies of molecular conformations.

## 2.9  Instrumentation in fluorimetry

The optical requirements of a fluorescence spectrometer are quite different from those of an absorption spectrometer. Therefore the use of 'fluorescence attachments' for UV–visible spectrometers is not usually successful. The characteristic right-angled optics of a fluorescence instrument (Fig. 2.8) allow the fluorescence

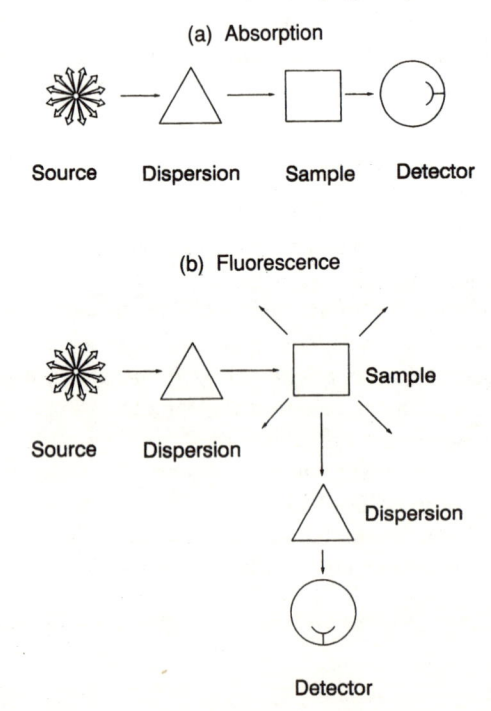

Fig. 2.8  *Optical layout of (a) absorptiometry and (b) fluorimetry.*

to be detected in the absence of the transmitted beam; they also minimize (but by no means eliminate) STT Rayleigh scattered light. Since the wavelength of the emitted light is longer than that of the absorbed light, two monochromators (or filters) are also needed, giving rise to several types of spectrum. Although the excitation beam passes through a filter or monochromator before striking the sample, photodecomposition is a serious problem in fluorimetry because of the powerful light sources used. Most instruments therefore incorporate a sample compartment shutter which protects the sample until the moment of measurement.

Almost all fluorimeters are effectively single beam instruments, although a secondary beam is often used in the excitation optics for 'ratio recording' and spectral correction. A small portion of the light from the excitation monochromator is passed to the surface of a 'quantum counter', i.e. a fluorescence emitter whose output is proportional to the number of incident photons, irrespective of their wavelength. The signal from the quantum counter is monitored by a secondary detector whose output can be compared with that of the main detector. Properly designed and used, such a system corrects for both temporal variations in light source intensity (for example during the collection of a spectrum), and for the wavelength dependence of the light source output plus excitation monochromator characteristics. The latter wavelength-dependent effects often distort the excitation spectrum of a compound substantially and as a result it looks very different from the UV absorption spectrum of the same compound.

The fluorescence intensity from a given sample is proportional to the incident light source intensity. Fluorimetry thus requires bright but stable sources. If excitation spectral information is to be obtained a broad-band output from the source is also necessary. Although quartz–halogen lamps have been used successfully for visible region fluorimetry, many fluorophores absorb in the ultraviolet region so two types of light source have become commonplace. Mercury lamps emit essentially line spectra (so spectral scans are not possible) but they operate conveniently in conjunction with filters in simple instruments. Moreover, they are stable and robust. The principal lines from a mercury lamp depend on its pressure: low pressure lamps produce light largely at 254 nm (very useful in biochemistry and for the detection of small aromatic molecules); higher pressure lamps have a good output at 365 nm, another most useful region. Scanning spectrometers almost all use some form of xenon lamp. The usable output spectrum extends from about 230 nm up to at least 900 nm, with the Xe atomic lines superimposed on the continuum, especially at 470–480 nm. A conventional 150 W Xe lamp (7.5 A, 20 V) requires a special power supply and also a high voltage starter circuit to strike the arc between the two electrodes which are about 2 mm apart. The lamp is filled at ~3 bar pressure and operates at ~70 bar, thus necessitating stringent safety precautions. (The intense UV output is also, of course, highly dangerous to the eyes and skin). These lamps last a few hundred hours, their demise being signalled by a drastic reduction in UV output due to coating of the lamp envelope, and/or sudden intensity changes due to electrode damage and arc wander. Pulsed xenon lamps, which operate at an average power of only about 7.5 W, provide short (20 μs) but very bright (several kilowatt) light pulses, and generate less heat and less ozone. They require less specialized power supplies and find additional uses where pulsed source/time resolution spectrometry is needed. Their spectral profile is much the same as that of the continuous wave (CW) lamps (Fig. 2.2).

The exciting light is normally isolated either by an interference filter or by a grating monochromator (double monochromators have again been used on occasion). This grating is often blazed for maximum efficiency in the 250–300 nm region, and the monochromator (like its emission counterpart) can be driven at various speeds to scan spectra.

The sample compartment in a fluorescence spectrometer should have ample space for such accessories as polarizers, phosphoroscopes, and thermostatting equipment. The latter is particularly important as fluorescence intensities can be very temperature dependent. Decreases of up to 5% per degree increase in temperature have been described, although the average value is considerably lower. The use of intense light sources worsens the heating problem. Indeed, temperature control should be used much more often than it is at present. Fluorescence cells are normally polished on all four faces, and are usually fabricated from plastics or silica. Attempts to increase fluorescence signals by coating two adjacent faces with a reflective material have not proved successful, the increase in background scattering and stray light being at least as great as the gain in fluorescence intensity. Inner filter effects (optical artefacts) are very common in fluorimetry, and in some instruments a simple offset system is used to minimize these effects. Again a variety of microcells and flowcells are available.

The emission monochromator in a fluorescence spectrometer is usually a grating, although motor-driven 'interference wedges', i.e. continuously variable interference filters, have been used at wavelengths above 400 nm. Emission gratings are usually blazed for maximum efficiency at 450–500 nm. In terms of detection, photomultipliers are almost universally used. With all these optical elements, the wavelength dependence of their efficiencies can again produce spectral distortions, but correction procedures are available.

## 2.10 Fluorescence spectra and their correction

The availability of two scanning monochromators on many fluorescence spectrometers allows the use of several different kinds of spectrum to characterize any given fluorophore. If the excitation monochromator is scanned at a fixed emission wavelength, an 'excitation' spectrum is obtained. (Note that it is still fluorescence that is being detected!) This spectrum will, in a single beam instrument, reflect the wavelength dependence of the light source and excitation monochromator. Correction using a quantum counter should in most cases produce a spectrum identical with the conventional absorption spectrum of the same compound.

Scanning the emission monochromator at a fixed excitation wavelength generates the 'emission' or 'fluorescence' spectrum. This also will be uncorrected unless special precautions are taken. It will reflect the wavelength dependence of the properties of the emission monochromator and of the detector sensitivity. Emission spectra can be corrected in several ways, including the use of the pre-calibrated light source to obtain emission correction factors (this approach multiplies errors); the use of a separate, calibrated external light source to generate the correction factors; or the use of a known, standard emission spectrum to produce the factors. Modern computer-linked spectrometers achieve such corrections most readily. Only when fully corrected spectra are used can they be expected to be the same on different instruments, or even on the same instrument on different occasions.

It is clear that the full description of the fluorescence properties of a compound requires a three-dimensional presentation, the axes being the excitation and emission wavelengths and the observed intensity. Such 'excitation-emission matrix' or EEM spectra are available either as isometric projections or as contour diagrams (Fig. 2.9). On a conventional spectrometer they have to be obtained by repeated scans, and with computer manipulation of the data the whole process can be quite tedious. A little thought shows that vertical and horizontal sections through the contour diagram represent emission and excitation spectra, respectively. A 45° section through the contours with the two wavelengths equal represents the Rayleigh scattered light component.

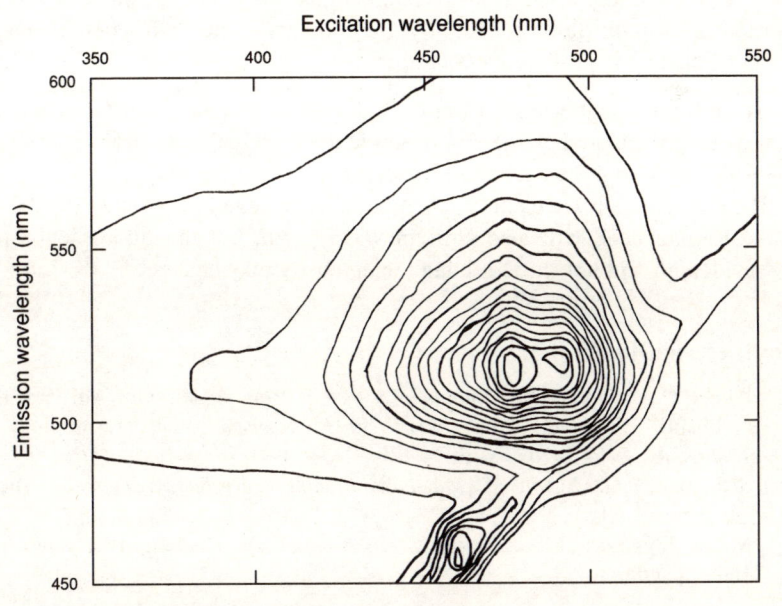

Fig. 2.9 *Contour spectrum of emission and excitation.*

A 45° section through the contours when the emission and excitation wavelengths are *not* the same corresponds to a so-called synchronous spectrum. This is obtained in practice by driving both monochromators at the same speed simultaneously with the emission wavelength greater by a fixed interval, $\Delta\lambda$. Synchronous spectra tend to be narrower than conventional excitation and emission spectra, thus facilitating the resolution of mixtures, and they have found numerous applications in organic and inorganic analysis. It is evident that, if the monochromators can be driven separately and at different scan rates, sections through the contours at angles other than 45° are obtained. These so-called 'variable angle' spectra also have attracted some interest.

## 2.11 General methods in fluorescence spectroscopy

The sensitivity of fluorimetry derives from the fact that in the limit of trace analysis, the technique detects a small emission signal above a zero background, rather than a small difference between two large numbers, as in absorptiometry. It is therefore essential to keep this background at a low level if possible and the initial precaution is the extreme cleanliness and purity of materials. Solutions should be free of suspended particles (which increase scattered light) and solvents should be tested to ensure that they do not contribute a background signal (for example water distilled three times from silica). In ultra-trace analysis, the signal from the cell itself may be the limiting factor. Plastic containers should not be used for water and other solvents.

A further source of background signal is the Raman scattering from the solvent. This occurs at longer wavelengths than the exciting light and can thus be confused with fluorescence by the unwary. The two phenomena are distinguished by the following conditions:

1. Since Raman scattering is from the solvent, its frequency shift and hence its wavelength are predictable. For example the principal water Raman band is at ~3400 cm$^{-1}$ from the excitation wavelength.
2. If the excitation wavelength is deliberately changed by around 10 nm, then the Raman peak will also shift in wavelength, but the fluorescence peak wavelength will not shift, although its intensity may change.

### Further reading

1. Parker, C.A. (ed.) (1968) *Photoluminescence of Solutions*, Elsevier, Amsterdam.
2. Guilbault, G.G. (ed.) (1991) *Practical Fluorescence: Theory, Methods and Techniques*, Dekker, New York.
3. Schulman, S.G. (ed.) (1985) *Molecular Luminescence Spectroscopy*, **1**, Wiley & Sons, New York.
4. Miller, J.N. (ed.) (1981) *Standards in Fluorescence Spectroscopy*, Chapman and Hall, London.

# 3 Standards in UV spectroscopy

## 3.1 Introduction

Long ago, but not quite so long ago that none of us can remember, or at least know someone who can remember, spectrometers were instruments of a very basic nature, rather primitively constructed of phototubes and moving spot galvanometers. In those days, the spectroscopist's results would have been totally meaningless if the instrument had not been carefully calibrated before the measurements, or afterwards, or preferably both.

At present, so manufacturers would have us believe, many instruments are self-calibrating and we can therefore forget the need for manual calibration. Or can we? Let us suppose that we've taken out measurements and they just don't look right. As good analysts, we go through the whole procedure again and get virtually the same results. What do we do then? Do we convince ourselves that the instrument must be right. After all, the manufacturers claim it is totally reliable and therefore the error must lie elsewhere, perhaps with the sample or with our technique. What else can we do?

The truth is that we still need to calibrate and to do this we need standards, perhaps even more so now than we did with the first instruments. What happens when we don't check our instrument calibration regularly is demonstrated in Table 3.1, which shows the results of one collaborative exercise. Each point represents the absorbance reading at 287nm on one instrument and the letters represent different types of instrument. Note that the absorbances measured vary from 0.040 to 0.360.

In the years since the Ultraviolet Spectrometry Group was founded (1948), spectrometers have undoubtedly become much easier to use. Consequently the analyst can use an instrument and get good results without knowing very much of what goes on inside the black box. But there is a real danger in accepting results blindly without enquiring into their validity and without having developed that sixth sense which tells the experienced analyst that there is something wrong in the results generated.

In one of the most entertaining books on spectrometry written by Edisbury, (a founder member of the UVSG), he referred to standards as his 'link with sanity'. In Edisbury's terms, a link with sanity was what was needed when, having doubted the validity of a reading and having eliminated all the obvious elementary things which could be responsible, we are still left with an anomalous value.

**Table 3.1** *Variation of absorbance readings produced by several calibrated spectrophotometers*

| A287nm | Spectrophotometer type |
|---|---|
| 0.400 | |
| 0.380 | |
| 0.360 | AC |
| 0.340 | CDDHK |
| 0.320 | CCDEEK |
| 0.300 | ACEFFG |
| 0.280 | AAFK |
| 0.260 | ABCCEF |
| 0.240 | AAABCCCCCCCCCCCCCCCF |
| 0.220 | ACCCCEEEEEFI? |
| 0.200 | AACCCEEEFIIJ |
| 0.180 | AAAAABDEEFFIII |
| 0.160 | AACCCCDDFFFFHHI? |
| 0.140 | ABBBCH |
| 0.120 | ACFFK |
| 0.100 | BCFF |
| 0.080 | EF |
| 0.060 | C |
| 0.040 | F |
| 0.020 | |
| 0.000 | |

Edisbury had a preference for solid standards and wrote[1]:

The essential feature of a link with sanity is immediate availability. No one is likely to use it, except in extremis, if more than a moment's preparation is needed.

Up to a point I agree with him. But solids only test the intrinsic performance of the instrument, whereas liquid standards, correctly used, can check the entire working procedure. However there is room for both modes and thus, let us first examine the pros and cons of liquid and solid standards (Table 3.2).

What then ought we to check? In most cases we are measuring an absorbance or transmission, therefore we need to check the absorbance/transmission scale for accuracy. For most peaks the absorbance varies considerably with wavelength. If the peak is sharp, any error in wavelength will result in a very considerable error in absorbance measurement. This becomes even worse if the measurement has to be taken on the upslope or downslope of the absorption band. That is why we try always to take our measurements at the plateau of a peak. Consequently we ought first to check the wavelength calibration. Then there is one more feature which should be considered. If there is likely to be a degree of stray light in the instrument and this is a common occurrence, then the absorbance measurements we

**Table 3.2** *Pros and cons of liquid and solid standards*

| Solid standards | Liquid standards |
|---|---|
| *Advantages* | *Advantages* |
| 1. Ease of use | 1. They check out the whole system |
| 2. No preparation required | 2. Absorbance levels easily vary |
| 3. Fairly insensitive to temperature | 3. Can be used as a 'blank' |
| 4. Stability | 4. Any cuvette shape or type (including flow systems) |
|  | 5. Homogeneous |
| *Disadvantages* | *Disadvantages* |
| 1. Only check part of the system | 1. Stability |
| 2. Absorbance level fixed | 2. Preparation |
| 3. May not be homogeneous across the field | 3. Temperature effects |
| 4. Stability cannot be assumed |  |
| 5. Fixed shape |  |
| 6. Must be kept clean |  |

make could be seriously in error, and we should have some method of determining stray light.

## 3.2 Wavelength standards

The ideal wavelength standard would have very narrow and very well defined peaks, in order that there is little doubt what the true wavelength should be and minimal uncertainty when we are sitting on the peak maximum (Fig. 3.1). The nearest we can get to this are the lines from an emission source (Fig. 2.2). The most convenient of these sources are hydrogen, deuterium or mercury, though many others are equally suitable. In fact many of the modern, so-called self-calibrating instruments make use of deuterium emission lines to set the wavelength scales (usually the lines at 486.0 or 656.1 nm). That does not mean

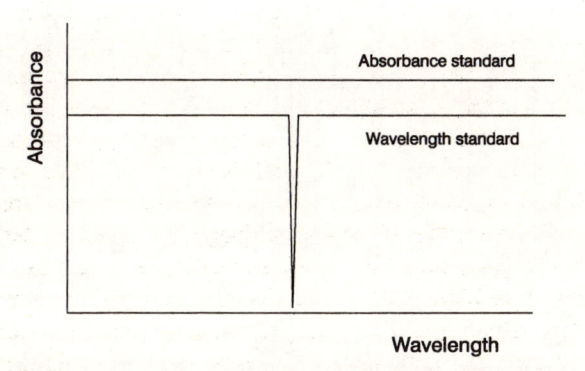

Fig. 3.1 *Ideal spectrum from a wavelength and absorbance standard.*

## 38  Standards in UV spectroscopy

these instruments are always correctly calibrated and in our laboratory we have experienced a software fault which set the calibration fairly precisely, but about 20 nm from where it should have been. However, since most routine spectrometric analysis is performed on rather broad bands, such a shift is usually not immediately obvious in the results obtained.

There are, however, instruments where the use of deuterium lines for this purpose is not available, therefore what other options are there? An older generation may have modified the instrument so as to make it possible to measure these lines, but it is unrealistic to expect that today with the current microprocessor designed instruments. One of the most convenient alternatives is to use a rare earth glass as a standard filter. This can be made of silica fused together with a rare earth oxide. The most commonly used are holmium, which gives absorption bands between 241 and 637 nm, and didymium (573 to 803 nm). Typical spectra from these filters are shown in Figs 3.2 and 3.3. The bands are generally fairly sharp, and if the glass is cut to size and set into a holder of the same dimensions as a 10 mm cuvette, the filter is very easy to use (Fig. 3.4). Such a filter clearly meets Edisbury's criterion of immediate availability. Unfortunately these glasses suffer from one major disadvantage: they are not totally reproducible. The wavelengths do vary somewhat from

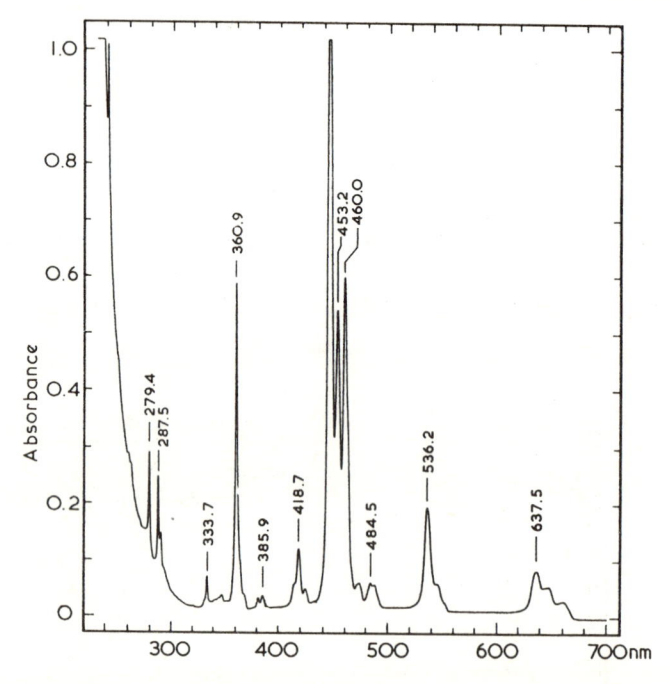

Fig. 3.2  *Absorption spectrum of a holmium filter of unknown origin, scanned at $1 n\ ms^{-1}$ with SSW = 0.2. The peak values are taken from McNierney and Slavin [2]. Note that there are variations in peak positions between different batches of glass.*

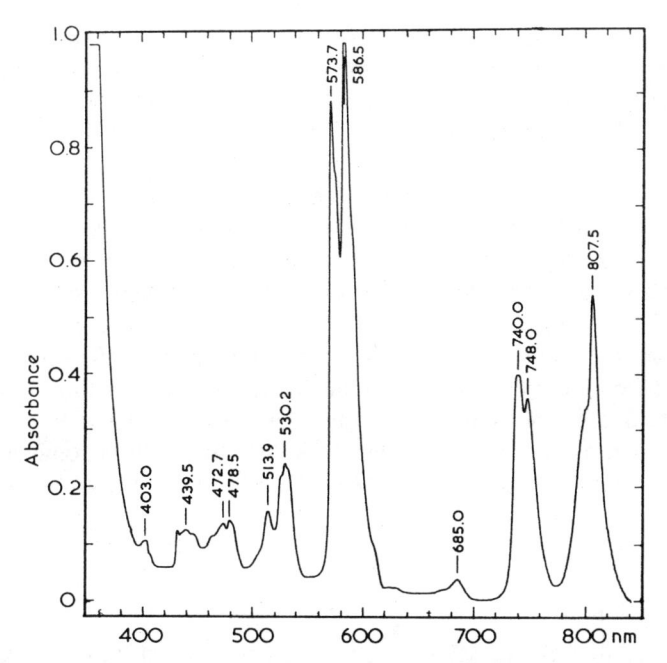

Fig. 3.3  *Absorption spectrum of a didymium glass filter (Chance-Pilkington ON 12, 2.0 mm thick), scanned at 1nms$^{-1}$ with SSW = 0.2. The peak values are taken from Perkampus et al. [3]. Note that there are variations in peak positions between different batches of glass.*

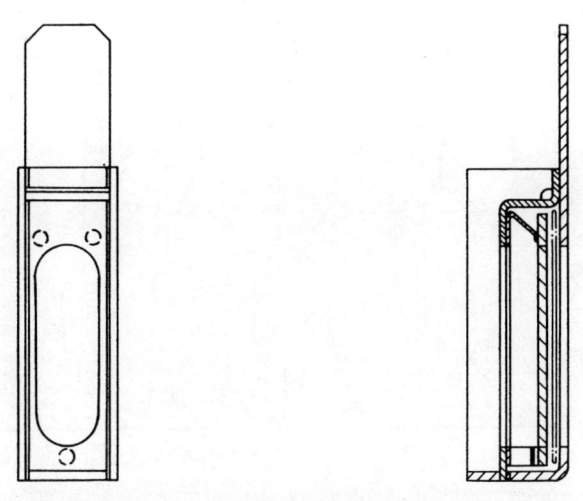

Fig. 3.4  *A stress-free mounting for glass filters, from Sharp [4].*

batch to batch; a batch may not always be completely homogeneous and the peak positions may be sensitive to slit width. But they are certainly satisfactory for instruments with spectral slit widths above 5 nm, and can be used to check day-to-day reproducibility of other instruments if the conditions used are rigorously controlled in order that they are always the same.

A better but less convenient approach is to use aqueous solutions of rare earth salts. These, if adequately mixed, do not suffer from the variability problems of the glasses, and as a bonus they generally have somewhat sharper absorbance bands. The solution most commonly used, probably because it is commercially available and recommended in the pharmacopoeias, is holmium perchlorate in dilute perchloric acid. Also available commercially are sealed cuvettes containing samarium perchlorate in perchloric acid. The spectra from these solutions are shown in Figs. 3.5 and 3.6.

More recently a new crystalline solid material has become available which has very sharp bands. It was developed by McCrone Research

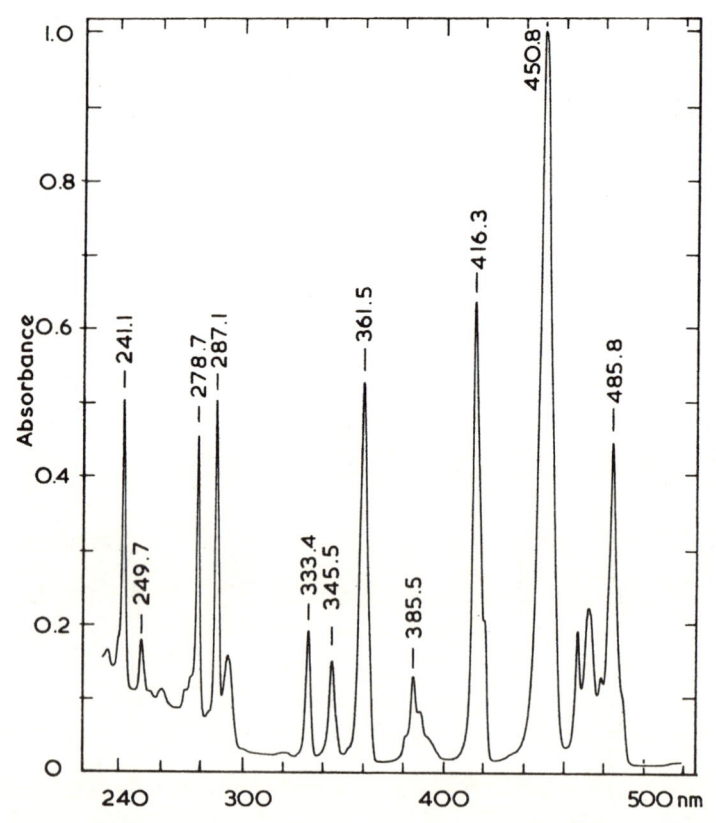

Fig. 3.5 *Absorption spectrum of holmium (III) perchlorate solution (10% w/v in 17.5% w/v perchloric acid) in a 10mm cell measured against a perchloric acid blank. Scanned at 1nms$^{-1}$ with SSW = 1.0nm. The peaks at 241.1, 287.1 and 361.5nm are recommended by the European Pharmacopoeia [5] for wavelength calibration. Taken from Burgess [6].*

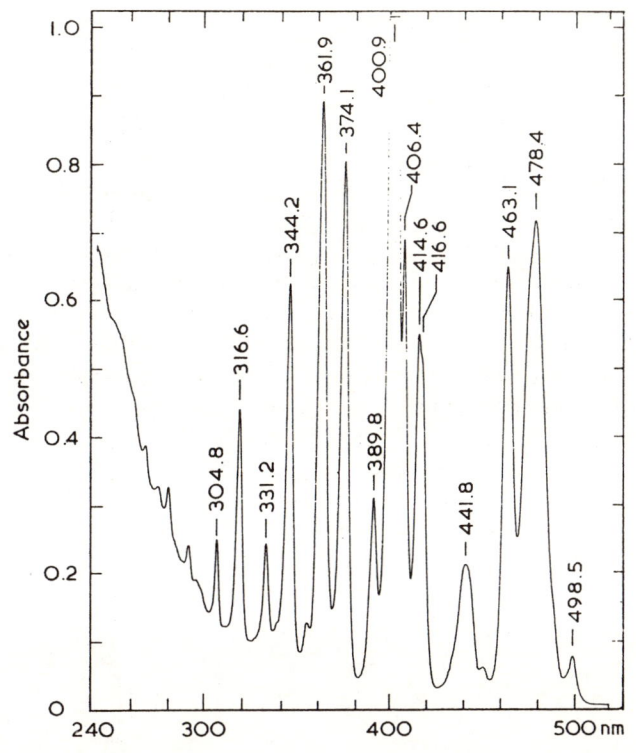

Fig. 3.6 *An absorption spectrum of samarium (III) perchlorate solution (10% w/v 17.5% w/v perchloric acid) in a 10mm cell measured against a perchloric acid blank. Scanned at 1 nm s$^{-1}$ with SSW = 1.0nm. Taken from Burgess [6].*

Associates and is available with certificate, from the National Physical Laboratory (Teddington, Middlesex, UK). It is based on Nd YAG (neodymium yttrium aluminium garnet) which was produced in the first instance for laser work. A typical spectrum for this is given in Fig. 3.7.

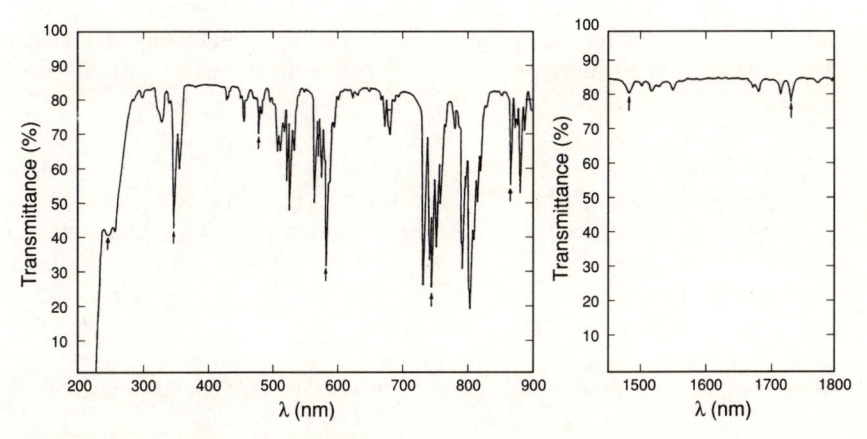

Fig. 3.7 *McCrone wavelength standard.*

## 3.3  Absorbance standards

Adopting the same approach as for wavelength standards, the ideal absorbance standard would be a substance whose absorbance is constant with wavelength, as in Fig. 3.1. If we had a series of solid substances with these characteristics, or could prepare a series of solutions at various concentrations, we could check our absorbance scale at all wavelengths. Unfortunately at present this type of material is not available.

The properties required of a solid absorbance standard have been listed by Irish (Unicam, Cambridge, UK). Ideally the standard should:

1. be convenient to use and simulate the normal use of the instrument;
2. have an absorbance independent of the wavelength setting;
3. be unaffected by stray light;
4. be non-fluorescent;
5. show little change in its optical properties with temperature;
6. not be changed by exposure to normal atmospheres and light; and
7. be easy to construct and calibrate.

For liquid standards we need to add that it should:

8. be insensitive to pH;
9. be highly pure and well characterised with a generally accepted molar extinction; and
10. obey Beer's Law.

Regrettably, again, this ideal standard does not exist.

Numerous solid standards have been advocated over the years and one approach has been the mechanical one of simply blocking part of the beam. But since the beam is unlikely to be homogeneous a rotating sector procedure was developed. However, rotating sectors are clumsy and may have other problems but they do have the unique advantage of giving absolute results, in that the relative areas of the disc can be measured mechanically. Apertures, wire meshes, and perforated screens have also been used, but are satisfactory only if the greatest care is taken to keep them clean and uncontaminated, otherwise they are prone to produce highly dubious results.

Glass filters of various types have been used with considerable success, but cannot themselves give absolute results and therefore must first be calibrated. Examples are the NBS glass filters (Fig. 3.8) which are of course supplied with an NBS certificate. These are coloured and the glass absorbs strongly below 350 nm and so are somewhat remote from our ideal. Better in this respect are the so-called 'neutral' glasses such as Chance ON10 (Fig. 3.9). Glass filters must be carefully polished in order to ensure that they are as near parallel as possible. At present the best glass available is Schott NG-4 (Fig. 3.10) and this is available both from the NBS in America and the

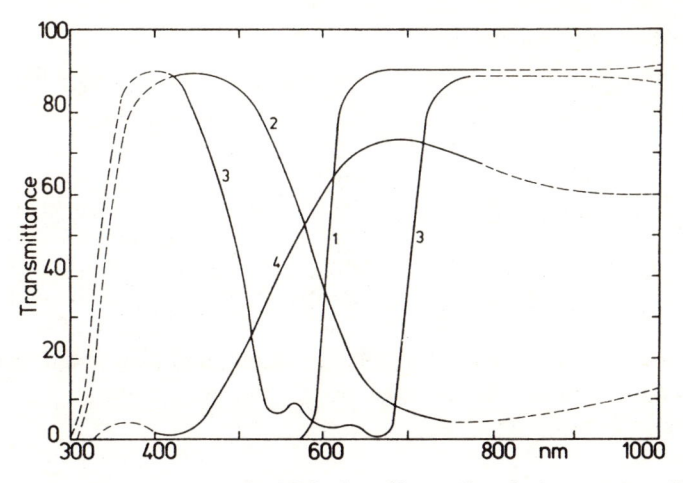

Fig. 3.8 *Transmittance spectra for NBS glass filters: (1) selenium orange; (2) copper green; (3) cobalt blue; (4) carbon yellow. From Gibson [7].*

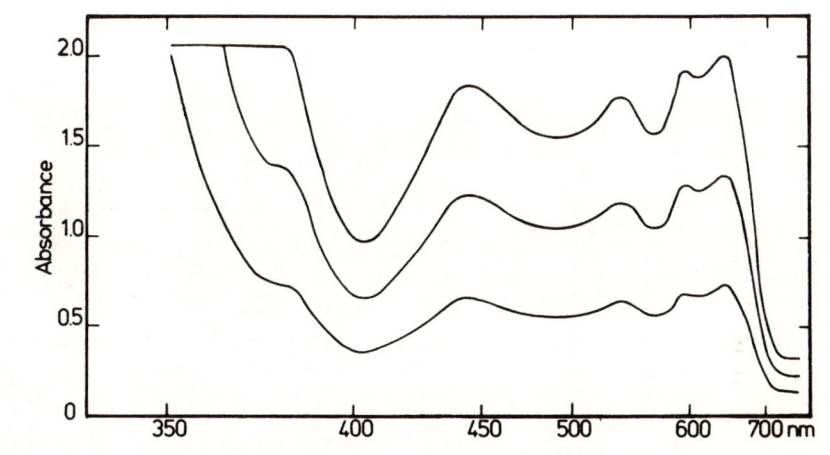

Fig. 3.9 *Absorption spectra of Chance ON10 filters of various thicknesses. From Slavin [8].*

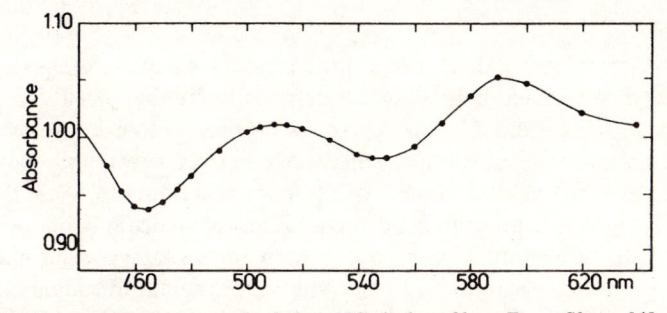

Fig. 3.10 *Absorption spectrum of a Schott NG-4 glass filter. From Sharp [4].*

NPL in the UK. The latter, as the result of a collaborative project with Unicam, offers a set of six filters, each in a special mounting, with nominal absorbances from 0.1 to 1.9.

Glass filters cut off in the UV region, therefore to fill this gap both NBS and NPL now offer filters made from quartz coated with an evaporated metal film. Typical curves for the NPL filters are shown in Fig. 3.11, and probably represent the closest approach yet to our ideal. They do however suffer from reflectance errors which depend primarily on the geometry of the incident radiation. For this reason their suitability ought to be investigated for the instruments in question before they are relied upon. In addition there are indications that they are not completely stable, and as a result should be returned for recalibration at intervals.

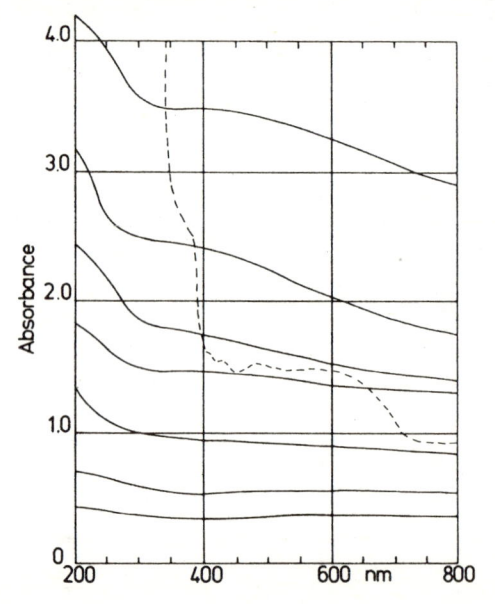

Fig. 3.11  *Typical absorbance spectra for NPL nichrome on fused silica filters (———) compared with a Schott NG-3 filter (------). From Clarke [9].*

As I have already stated, filters are easy to use but only check out the photometry. As a practicing analyst, I personally feel happier if I can check the whole of the system. Consequently I would prefer to use a solution even if it is in some ways less convenient. Numerous substances have been suggested over the years and most of these are listed in Table 3.3. The most commonly used is potassium dichromate. As a substance it is of course well characterised, available in very pure grades and spectral data have been reported on in acid, neutral and basic solution. Its main disadvantage is its reactivity as a powerful oxidizing agent. What this means in practice is that

**Table 3.3** *Suggested substances for liquid filters*

| 200–400 nm | 400–800 nm | 200–800 nm |
|---|---|---|
| $K_2Cr_2O_7$ –$H^+$ | Thomsons solutions | Dyes |
| $K_2Cr_2O_7$ –$OH^-$ | Cobalt (II) | Green food colours Frenchs' |
|  |  | Langdales sap green |
| $K_2Cr_2O_7$ –'neutral' | Ammonium cobalt (II) sulphate | Thomson's (modified) |
| $KCrO_4$ –$Na_2HPO_4$ | Nickel (II) |  |
| $KNO_3$ | Copper (II) |  |
| Pyrene | Iron – dipyridyl |  |
| Potassium hydrogen phthalate |  |  |
| Picric acid |  |  |
| Nicotinic acid |  |  |
| Salicylaldehyde |  |  |
| Anthraquinone |  |  |
| Acridine |  |  |
| Caffeine |  |  |

solutions have to be prepared with care, in clean apparatus, and discarded after use. This is a long way from Edisbury's 'immediate availability', but nevertheless must be regarded as excellent practise in the preparation of a standard. Unfortunately it is also markedly pH sensitive, forming an equilibrium system between dichromate in acid solution and chromate in alkali. However, in dilute acid, 0.01 N sulphuric or 0.001 N perchloric it gives a nice smooth curve with two maxima and two minima (Fig. 3.12) and

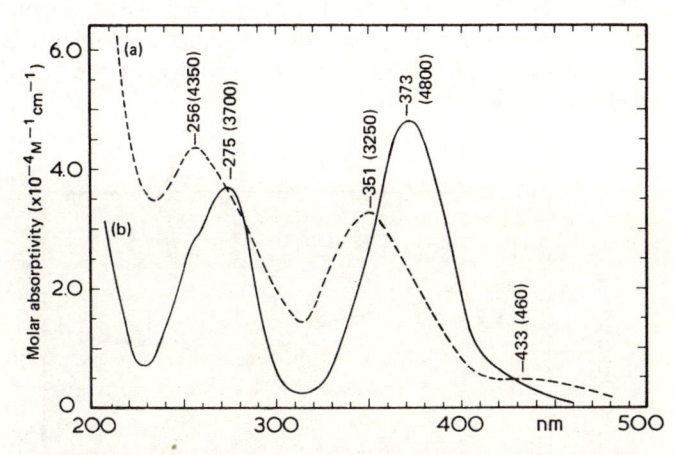

Fig. 3.12 *Absorbance spectra of (a) potassium dichromate in acidic solution (-----); and (b) potassium chromate in 5 mM potassium hydroxide (———). The latter is effectively identical with the spectrum of potassium dichromate measured under the same conditions. Redrawn from Perkampus et al.* [3].

is very easy to use. But as long as the solution is defined precisely and prepared correctly, most of the objections are not founded.

A group of substances advocated in some quarters are the solutions of cobalt, nickel and copper salts. Solutions of the nitrates are now made by the French Standards organisation Laboratoire National d'Essais (LNE). The National Bureau of Standards (NBS) also supply in sets with nominal absorbances of 0.3, 0.6, and 0.9. They give suitable peaks in the visible region, but they are rather narrow and consequently accurate wavelength setting or measurement at the peak is essential (Fig. 3.13).

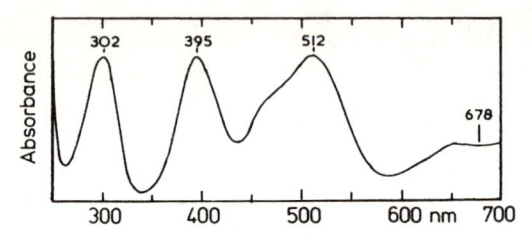

Fig. 3.13  *Absorbance spectrum of NBS Standard Reference Material 931. A set of three solutions is supplied, of maximum absorbance approximately 0.3, 0.6 and 0.9. From Burke et al. [10].*

Another approach was put forward by Thomson, as a neutral grey solution, contains salts of chromium (III and VI), cobalt (II) and copper (II) and needs to be prepared a few weeks before it is required, since it ages quite noticeably during the first few days, turning from pink to grey. Clearly, chemical changes are occurring. The end product has the spectrum shown in Fig. 3.14. A consequence of these changes is that it cannot provide an absolute standard, but does make a suitable transfer standard. It is also temperature sensitive and therefore should always be used under thermostatted conditions.

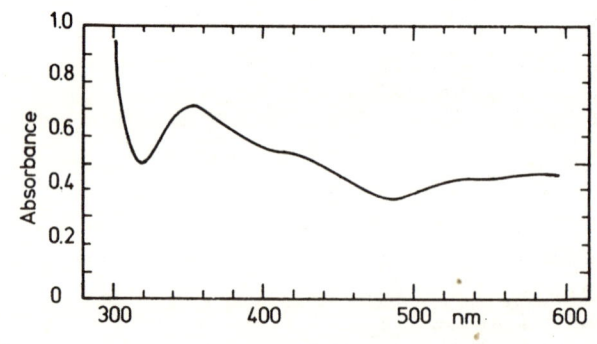

Fig. 3.14  *Absorption spectrum of Thomson's solution made up, aged, and measured in a 10mm cell. Redrawn from [11].*

Thomson's solution has another disadvantage in having a cut-off point at about 300 nm. The American National Bureau of Standards in Washington has proposed a modified Thomson's Solution containing chromium (III), potassium sulphate, cobalt (II) ammonium sulphate and 4-nitrophenol. This is transparent down to 230nm, but little work seems to have been carried out on the use of this solution as a standard since the original publication (Fig. 3.15).

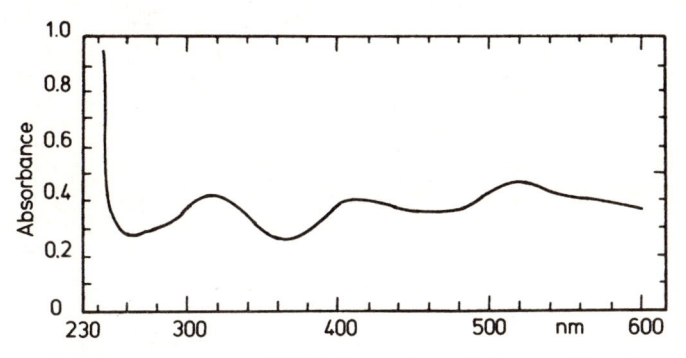

Fig. 3.15 *Absorption spectrum of NBS Composite solution, pathlength unspecified. From Menis and Schultz [11].*

## 3.4 Stray light

Stray light has been described as an insidious problem because you often do not know of its occurrence until you have cause to suspect your results. Although even then it is difficult to measure with any accuracy. What is it? Stray light is radiation of wavelengths other than the one you're working with, which nevertheless reaches the detector. It arises mainly from random reflections and scattering from the internal parts of the instrument. There is also a minor component due to the finite width of the slits allowing radiation just above and below the nominal to get through.

With a modern, well-maintained instrument, stray light should not be a problem. With older instruments, and those which have not been regularly serviced, it may be quite significant a problem. Its effect will, of course, depend markedly on the type of spectrum being run, but usually it results in an apparent absorbance somewhat lower than the true absorbance. At any particular wavelength, the stray light in an instrument will, to a first approximation, be independent of the sample. Hence the more strongly absorbant the sample, the greater the influence of stray light. This is shown graphically in Fig. 3.16. In effect, stray light limits the maximum absorbance you can measure. Consequently, if your sample absorbs most of the available incident radiation, you are quite likely to see the effects of stray light. If it does not and your measured absorbance is low, stray light is likely to be insignificant except, perhaps, where the lamp output is low as in the far UV. An example of the effects of stray light on a spectrum is shown in Fig. 3.17.

48  *Standards in UV spectroscopy*

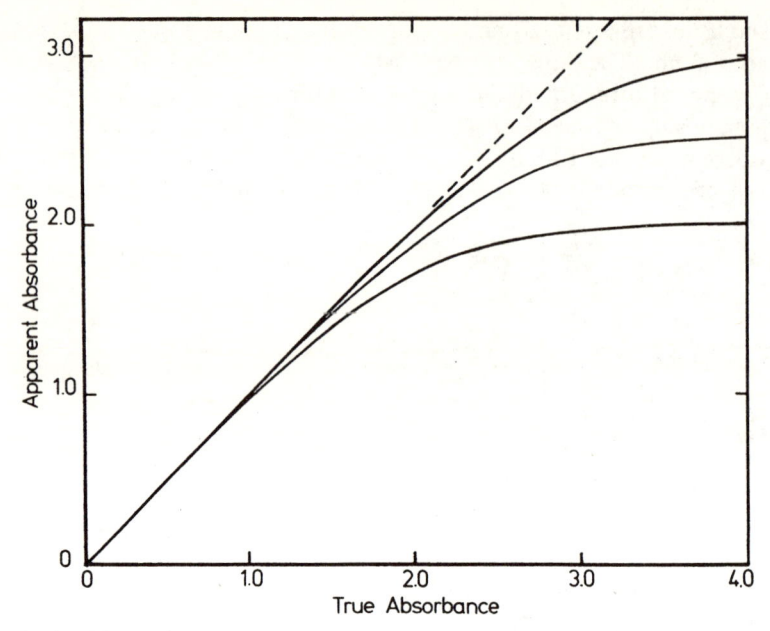

Fig. 3.16 *Plots of apparent absorbance against true absorbance for different percentages of stray light.*

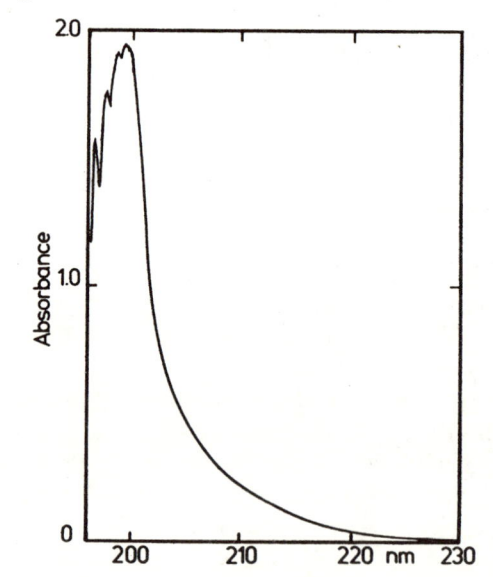

Fig. 3.17 *Absorption spectrum of ethanol showing the effect of stray light.*

As stray light varies with wavelength, another consequence can be a wavelength shift. We can demonstrate this effect by considering the measurement of maleic acid in ethanol at a range of concentrations as shown in Fig. 3.18. Under the same conditions, maleic acid in water does not show this effect, because the molar extinction, and consequently the absorbance, is lower (Fig. 3.19).

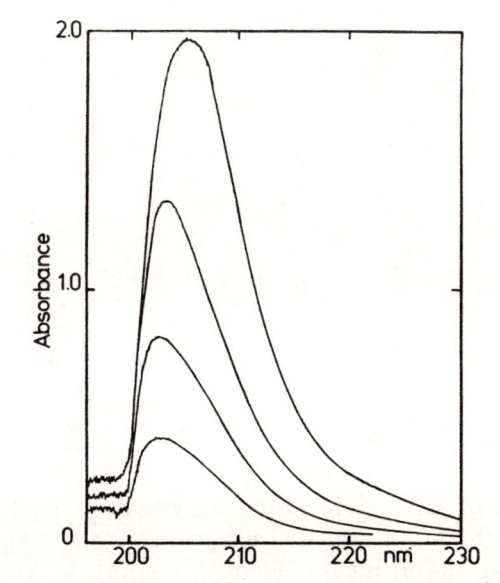

Fig. 3.18 *Absorption spectrum of maleic acid in ethanol at various concentrations, showing the effect of stray light.*

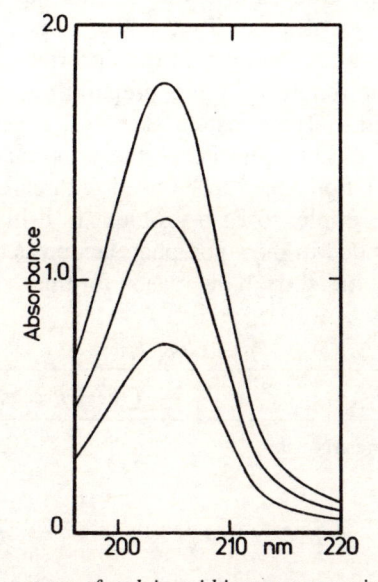

Fig. 3.19 *Absorption spectrum of maleic acid in water at various concentrations.*

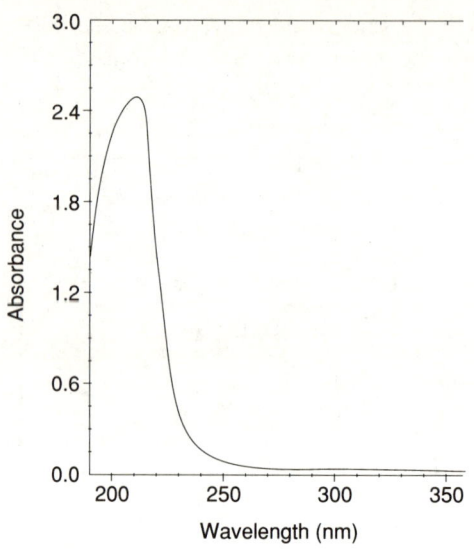

Fig. 3.20  *Spectrum of Vycor filter.*

Measurement of stray light with any precision is difficult, but an approximate determination which is adequate for most purposes, is readily obtained by using cut-off filters. Clearly, if you insert a cut-off filter i.e. a filter with a very high absorbance and sharp profile into the beam, then any light transmitted must be of wavelength higher or lower than the cut-off. The only solid material which seems to have been generally used for this purpose is Vycor. This absorbs at about 205 nm but its absorption profile leaves something to be desired (Fig. 3.20). It is nevertheless a useful material for a quick check, and meets the criterion of being immediately available. Better, but requiring some preparation, are salt solutions of various types as listed in Table 3.4. In all cases, these should be used just inside the cut-off edge, say 5 nm, hence the importance of a sharp cut-off (Fig. 3.21). These solutions need to be used with care and common sense. There appears, for example, to be a problem in using potassium chloride solution as recommended in the some pharmacopieas, since the absorbance drifts markedly with time, though the reason for this is not known.

**Table 3.4**  *Cut-off filters for stray light tests*

| Salt solution | Filter wavelength range (nm) |
| --- | --- |
| Saturated lithium carbonate | ~225 |
| 12 g dm$^{-3}$ KCl | 175–200 |
| 10 g dm$^{-3}$ NaBr | 195–223 |
| 10 g dm$^{-3}$ NaI | 210–259 |
| Acetone | 250–320 |
| 50 g dm$^{-3}$ NaNO$_2$ | 300–385 |

Recommended path length is 10 mm.

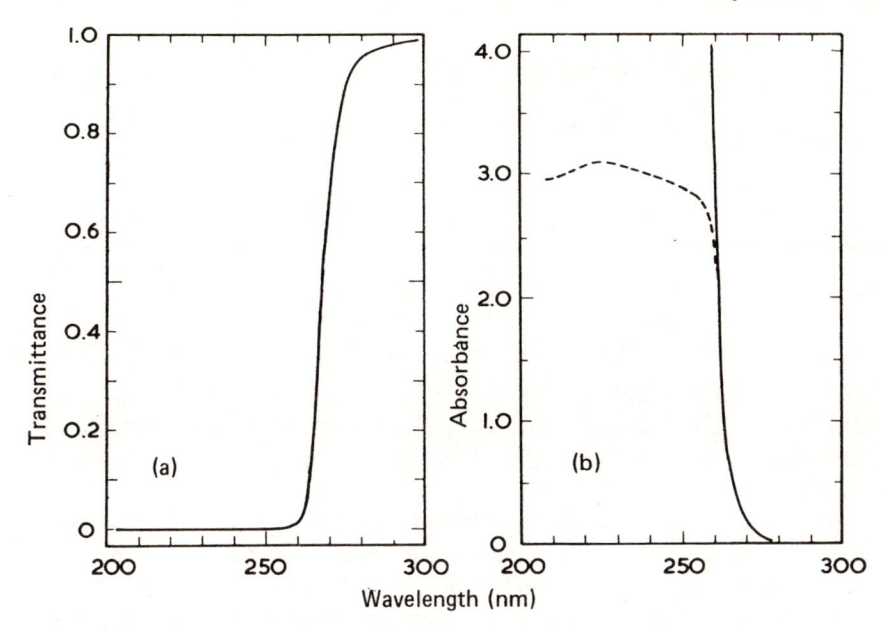

Fig 3.21 *Spectra of sodium iodide in water ($10g\ l^{-1}$) in a 10mm cell. (a) Transmission spectrum, (b) true absorption spectrum (—) and apparent absorbance (----). (From [12]).*

## References

1. Edisbury, J.R. (1966) *Practical Hints on Absorption Spectrometry*, Hilger and Watts, London, p. 173.
2. McNierney, J., and Slavin, W. (1962) *Appl. Optics*, **1**, 365.
3. Perkampus, H., Sandeman, I., and Timmons, C.J. (1971) *UV Atlas of Organic Compounds*, Butterworths, London/Verlag Chemie, Weinheim.
4. Sharpe, M.R. (1975) *UV Spectrom. Grp. Bull.*, **3**, 57.
5. *European Pharmacopoeia* (1980), 2nd edn, Section V.6.19, Maisonneuve, Sainte Ruffine, France.
6. Burgess, C. (1977) *UV Spectrom. Grp. Bull.*, **5**, 77.
7. Gibson, K.S. (1949) *NBS Circular* 484.
8. Slavin, W. (1962) *J. Opt. Soc. Amer.*, **52**, 1399.
9. Clarke, F.J.J. (1977) *UV Spectrom. Grp. Bull.*, **5**, 104.
10. Burke, R.W., Deardorff, E.R., and Menis, O. (1972), *J. Res. NBS*, **76A**, 469.
11. Menis, O., and Schultz, J.I. (1970), *NBS Technical Note* 544.
12. Knowles, A. and Burgess, C. (1984) *Practical Absorption Spectrometry*, Chapman and Hall, London, p. 216.

# 4 Spectroscopic multicomponent analysis

## 4.1 Introduction

The traditional approach to analysing mixtures focuses on the physical separation of each component prior to its measurement and quantification. Liquid chromatography is the separation technique and spectroscopy the preferred measurement technique.

Historically, prior to the advent of High Performance Liquid Chromatography (HPLC), UV spectroscopy was widely used to assay materials. However, the methodology was generally based on single wavelength measurements and its specificity was often very limited and highly application dependent. Often the analyte needed to be chemically or physically separated from the sample matrix to allow this simple spectroscopy to be used.

Those methodologies were rapidly overtaken by chromatographic techniques which were developed because of their powerful combination of separative and quantitative capabilities.

HPLC rapidly established itself in most industries as the preferred technique for analysis and for many applications UV detection was the measurement technique. Initially the potential of HPLC was limited by these detection systems based on simple single wavelength UV filter instruments. Subsequently these deficiencies were to be overcome by the introduction of variable wavelength UV instruments. These significantly increased the specificity of the measurements. Programmable UV detectors then followed allowing the detection of analytes to be optimized during a chromatographic run by wavelength stepping.

The next major step in the capability of HPLC came with the introduction of diode array spectrophotometers. These allow a full spectrum to be obtained in a fraction of a second. A full spectroscopic profile of an HPLC event can be obtained and a third dimension was introduced to chromatography, viz wavelength. This has proved invaluable in recognizing the presence of unresolved components in a chromatogram, identifying impurities, peak purity measurements, and HPLC optimisation.

As the power of the wavelength dimension of HPLC was investigated it became clear that UV could offer the analytical scientist an alternative technique for separating and determining the concentrations of analytes in a mixture by mathematical rather than chromatographic separation [1].

This chapter will focus on two relatively simple approaches to multi-component analysis using the full spectrum information and will introduce the more intensive techniques that are available based around multivariate statistical techniques.

## 4.2 Multivariate calibration techniques

Thoroughly investigating the quantitative information content of a spectrum is an exercise in multivariate data analysis. This information is spread throughout the wavelength range of the spectrum and each wavelength element or digitized data point can be considered as a separate variable [2].

In a typical UV–visible spectrum collected on a Hewlett Packard HP8452 diode array spectrophotometer, for example, there may be up to 321 data points. Other areas of the spectrum such as NIR or mid-IR may contain spectra of over 3500 data points each. For such applications it is necessary to use techniques which can reach into potentially large data sets and isolate the patterns in the data which represent their information content.

Multivariate calibration techniques come into their own when analysing the composition of complex mixtures, especially those involved in process control applications. Here the individual components may be spectroscopically similar or may interact with each other, and simple peak height or area calibration techniques are invalidated or difficult to implement.

## 4.3 Principles

### 4.3.1 *The Beer–Lambert law*

The fundamental relationship between UV spectral response and concentration is described by the Beer–Lambert law.

The absorbance of a measured absorption band is a function of the measurement wavelength, the thickness of the sample and the concentration of the absorbing species being measured. This is expressed as follows:

$$A_{1\lambda} = e_{1\lambda} c_1 b \quad (1)$$

where $A_{1\lambda}$ is the absorbance of species 1 at wavelength $\lambda$, $e_{1\lambda}$ is the absorptivity of species 1 at wavelength $\lambda$, $c_1$ is the concentration of the absorbing species 1, and $b$ is the optical thickness of the sample or pathlength of the measurement cell.

If a second absorbing species is present in the sample then, assuming no interactions between the components, the total measured absorbance at any wavelength will be the sum of the individual absorbances of the two species.

$$A_{\text{net}\,\lambda} = e_{1\lambda} c_1 b + e_{2\lambda} c_2 b \quad (2)$$

For an *n*-multicomponent case the relation becomes:

$$A_{\text{net}\,\lambda} = e_{1\lambda} c_1 b + e_{2\lambda} c_2 b + \ldots + e_{n\lambda} c_n b \quad (3)$$

and at constant unit pathlength

$$A_{\text{net}\lambda} = \sum_{\bar{n}=1}^{n} e_{\bar{n}\lambda} c_{\bar{n}} \qquad (4)$$

Similar expressions are written for other wavelengths in the spectrum and can be conveniently written in matrix form:

$$A = EC \qquad (5)$$

For the two-component situation discussed above, measured at three wavelengths we get:

$$A = \begin{vmatrix} e_{11}c_1 + e_{21}c_2 \\ e_{12}c_1 + e_{22}c_2 \\ e_{13}c_1 + e_{23}c_2 \end{vmatrix} \quad E = \begin{vmatrix} e_{11} & e_{21} \\ e_{12} & e_{22} \\ e_{13} & e_{23} \end{vmatrix} \quad C = \begin{vmatrix} c_1 \\ c_2 \end{vmatrix} \qquad (6)$$

Usually in an assay the absorptivities are not known and it is not necessary to measure them directly. A good estimate can be obtained by measuring the absorbance of standard solutions of the pure components at each wavelength. These are obtained from:

$$e_{n\lambda} = A^*_{n\lambda} / c^*_n \qquad (7)$$

where the symbol * denotes a standard solution.

Thus, for an analytical solution or an unknown, the values of *A* are known from the spectra and *E* from the spectra of the pure components we can estimate the values of *C*.

### 4.3.2 Simple least squares regression

Values of concentration, *C*, are most easily obtained by the method of least squares and some simple matrix operations accomplish this [3]. For example, using Equation 5,

$$A = EC$$

If this equation is pre-multiplied by *E'* it becomes:

$$E'A = E'EC$$

where *E'* is the transpose of *E*.
Then,

$$(E'E)^{-1}E'A = (E'E)^{-1}(E'E)C$$

Thus giving

$$C = (E'E)^{-1}E'A \qquad (8)$$

Simple estimates of the component concentration or proportions as in this case are not useful if there is no estimate of the error associated with them. A useful first estimate of error is obtained by looking at the difference or residual between the measured sample spectrum and the spectrum created mathematically from the calculated concentrations and the spectra of the pure components.

This synthetic spectrum is calculated from

$$B = EC \tag{9}$$

where $C$ are the calculated concentrations and $B$ the synthetic spectrum.

The residual spectrum, $R$ is given by

$$R = A - B \tag{10}$$

and the residual variance $S^2$ is expressed

$$S^2 = \frac{R'R}{(M-2)} \tag{11}$$

where $M$ is the number of wavelengths or data points used in the calculation.

The residual spectrum and residual variance are useful diagnostic tools for developing the multicomponent calculations. They allow the wavelength range used in the calculation to be optimized to minimise the residuals and hopefully maximize the accuracy.

Diode array spectrophotometers, in particular the Hewlett Packard HP8450, HP8451 and HP8452 series often calculate an estimate of the error in each absorbance value they measure. This error value at each data point can be used in the linear regression calculations as a weighting function allowing a better estimate of the calculated concentrations to be obtained. Absorbance values with low errors are emphasised more than those with high associated error values. This is weighted regression. In the absence of error values for each data point the diagonal elements of the covariance matrix $V$ are the variance estimates for each corresponding concentration and the square roots of these are the standard error estimates of the concentrations.

The covariance matrix $V$ is given by

$$V = (E'E)^{-1}S^2 \tag{12}$$

## 4.4  Simple linear regression: a tutorial

The uses of this approach can best be illustrated with a tutorial example. We shall consider a simple two-component mixture situation, that of the methyl orange pH driven equilibrium.

$$A \rightleftharpoons B$$

## 56  Spectroscopic multicomponent analysis

The UV spectra we will use are of methyl orange in solutions of different pH.

Shown in Fig. 4.1 are the UV spectra of methyl orange in solution at pH 2.2 and 5.2. These, we will assume, represent the extremes of the equilibrium and can therefore be treated as standards for A and B.

The spectra shown in Fig. 4.2 are treated as two component unknown mixtures and represent solutions of pH values between 2.6 and 4.6.

To simplify the calculations only 10 absorbance points are used in the calculations. These are taken at 30 nm intervals through the spectrum starting at 300 nm.

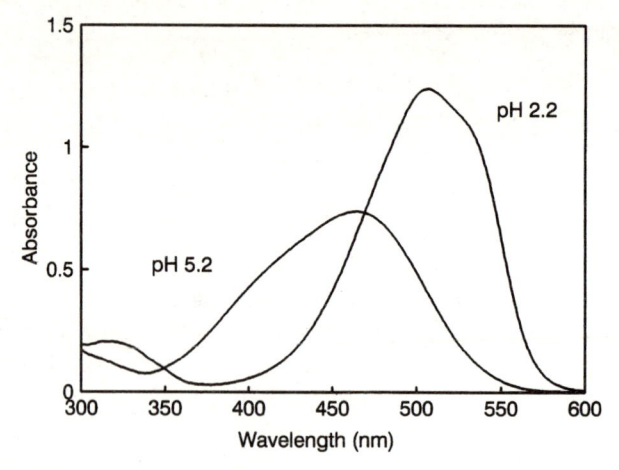

Fig. 4.1  *UV spectra (standards) of methyl orange.*

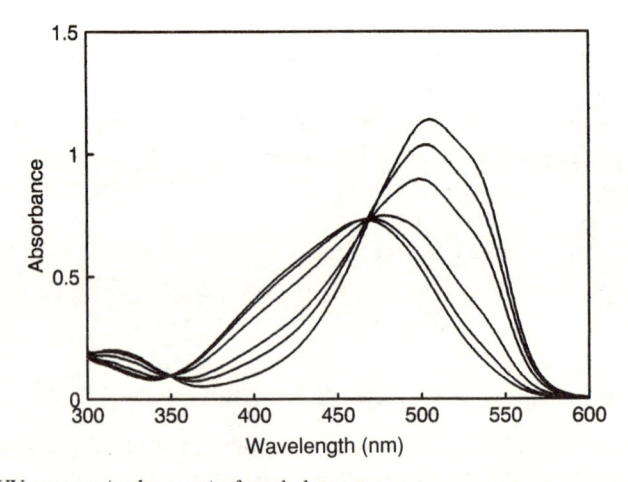

Fig. 4.2  *UV spectra (unknowns) of methyl orange.*

## Simple linear regression

In the example that will be discussed we are interested in the proportions of two species A and B in the mixture rather than their absolute concentrations. In this case we will assume that the concentrations of A and B in the standards are:

pH 2.2    [A] = 1    [B] = 0
pH 5.6    [A] = 0    [B] = 1

The absorptivity matrix is calculated from:

$$A = \begin{matrix} & \overbrace{\qquad pH \qquad}^{} & \\ & 2.2 \quad\;\; 5.2 & Wavelength\ (nm) \\ & \begin{vmatrix} 0.19081 & 0.23633 \\ 0.18817 & 0.13487 \\ 0.05862 & 0.16141 \\ 0.04079 & 0.33569 \\ 0.12001 & 0.52075 \\ 0.39102 & 0.66321 \\ 0.89333 & 0.65114 \\ 1.18384 & 0.35780 \\ 0.92107 & 0.09222 \\ 0.13020 & 0.00592 \end{vmatrix} & \begin{matrix} 300 \\ 330 \\ 360 \\ 390 \\ 420 \\ 450 \\ 480 \\ 510 \\ 540 \\ 570 \end{matrix} \end{matrix}$$

and therefore

$$E = A$$

and

$$E'E = \begin{vmatrix} 3.30905319 & 1.50642355 \\ 1.50642355 & 1.14435453 \end{vmatrix}$$

*Note*: For matrix manipulations it is important to carry out the calculations using as much precision as practical. This minimizes the effects of rounding errors, especially in matrix inversion. Final concentration values may then be truncated to 'sensible' proportions reflecting the accuracy of the standard and sample preparations.

Then,

$$(E'E)^{-1} = \begin{vmatrix} 0.56172438 & -0.57007596 \\ -0.57007596 & 1.25224521 \end{vmatrix}$$

For the remaining solutions which are to be treated as unknowns we create the following matrix:

## 58 Spectroscopic multicomponent analysis

$$A = \begin{array}{c} \overbrace{\phantom{aaaaaaaaaaaaaaaaaaaaaaaaaaaaaaaaaaaaaaaaaaaaaaaa}}^{pH} \\ \begin{array}{|cccccc|} 2.6 & 3.0 & 3.4 & 3.8 & 4.2 & 4.6 \\ 0.19273 & 0.20145 & 0.21107 & 0.22021 & 0.22569 & 0.22630 \\ 0.17799 & 0.17323 & 0.16882 & 0.15327 & 0.14339 & 0.13722 \\ 0.06348 & 0.07841 & 0.09416 & 0.12639 & 0.14653 & 0.15413 \\ 0.06969 & 0.11092 & 0.15271 & 0.24620 & 0.29723 & 0.31552 \\ 0.16057 & 0.21364 & 0.27225 & 0.39912 & 0.46883 & 0.49455 \\ 0.41476 & 0.44553 & 0.49207 & 0.58299 & 0.63268 & 0.64804 \\ 0.85446 & 0.80882 & 0.79367 & 0.73361 & 0.69809 & 0.67484 \\ 1.08066 & 0.95833 & 0.86766 & 0.63115 & 0.49246 & 0.42442 \\ 0.81969 & 0.70139 & 0.60507 & 0.36598 & 0.22501 & 0.15614 \\ 0.11275 & 0.09785 & 0.08319 & 0.05194 & 0.02933 & 0.01431 \end{array} \end{array}$$

and calculate $E'A$

$$E'A = \begin{vmatrix} 3.070596 & 2.795828 & 2.613195 & 2.110562 & 1.812101 & 1.654633 \\ 1.481176 & 1.462403 & 1.490764 & 1.507814 & 1.511528 & 1.495864 \end{vmatrix}$$

and, from this

$$B = (E'E)^{-1}(E'A)$$

$$B = \begin{array}{c} \overbrace{\phantom{aaaaaaaaaaaaaaaaaaaaaaaaaa}}^{pH} \\ \begin{array}{|cccccc|c} 2.6 & 3.0 & 3.4 & 3.8 & 4.2 & 4.6 & Conc. \\ 0.88 & 0.74 & 0.62 & 0.33 & 0.16 & 0.08 & A \\ 0.10 & 0.24 & 0.38 & 0.68 & 0.86 & 0.93 & B \end{array} \end{array}$$

The relative concentrations of A and B are truncated to two decimal places in keeping with the accuracy of the original experiment.

This simple least squares approach to multicomponent analysis is ideal as an introduction to the concepts of the technique. With a limited number of wavelengths the matrices can be processed using only a calculator and for many wavelengths they are ideal as a programming exercise.

### 4.5 The Kalman filter

Conventional approaches to multicomponent analysis try to fit the complete spectral shape of the sample spectrum to some combination of the standard spectra in one go, the relative proportions of each standard being adjusted until the best fit is obtained.

A Kalman filter achieves this end but does so by fitting the absorbances at each wavelength one at a time [4].

The basic Kalman filter operates using two models: the calculation or state model and the measurement model. The calculation model for multicomponent analysis is simple. It states that the concentrations of the components forming the sample spectrum at one wavelength are exactly the same at the next wavelength. This assumes the sample is not decomposing or reacting during the time it takes to collect a spectrum. The mathematical form of this model is given by:

$$C_\lambda = C_{\lambda-1} + Q \qquad (13)$$

where $C_\lambda$ is the concentration matrix of the components at wavelength 1, $C_{\lambda-1}$ is the concentration matrix at wavelength $\lambda-1$, and $Q$ is the model error.

The model error, $Q$, has properties similar to those of experimental error. It can be best thought as a corset. A small value of $Q$ means that the model must be adhered to strictly whereas a large $Q$ means that there is some flexibility. Usually $Q$ is given some small value and held constant through the calculation.

The measurement model relates the measured quantity, in this case absorbance, to the quantities required, in this case concentrations. In this example the equation of the model is Beer's law:

$$A_\lambda = E'_\lambda C_{\lambda-1} + R_\lambda \qquad (14)$$

where $A_\lambda$ is the absorbance at wavelength $\lambda$, $C_{\lambda-1}$ is the matrix of current concentrations for the components calculated for wavelength $\lambda-1$, $E'_\lambda$ is the transpose matrix of absorptivities of the standard solutions at wavelength $\lambda$, and $R_\lambda$ is the measurement error.

The detailed way in which the model and measurement equations are used is beyond the scope of this chapter and details are given in the references.

The difference between the sample absorbance and the calculated absorbance from current proportions of the standards is used to adjust the trial fit at the next wavelength. As the filter moves along the wavelengths in the spectrum the fit gets better and eventually converges on a stable ratio of the standards. Because of this mode of operation wavelengths may be presented to the filter in any order and convergence will still occur. However, as spectra are traditionally measured sequentially from high to low wavelengths most applications use this sequence.

This can be seen pictorially in Fig. 4.3 for a single component. At wavelength $\lambda_1$, we have an estimate of the component concentration $c(\lambda_1)$ which has been obtained from the absorbance $A(\lambda_1)$. This concentration is used at wavelength $\lambda_2$ with the response factor $e(\lambda_2)$ to calculate the absorbance $A(\lambda_2)'$ expected at wavelength $\lambda_2$. The actual absorbance is $A(\lambda_2)$ and the difference $A(\lambda_2) - A(\lambda_2)'$ is a measure of the bias in the estimate of the concentration. This bias is used to give a new estimate $c(\lambda_2)$ and the process is repeated for the next wavelength. This recursive estimate and update cycle leads to convergence on stable concentration estimates. An initial estimate of concentration is necessary to start off the filter and generally this is chosen to be $c = 0$.

The Kalman filter has been developed over a number of decades and applied in many areas, in particular those of electronic engineering.

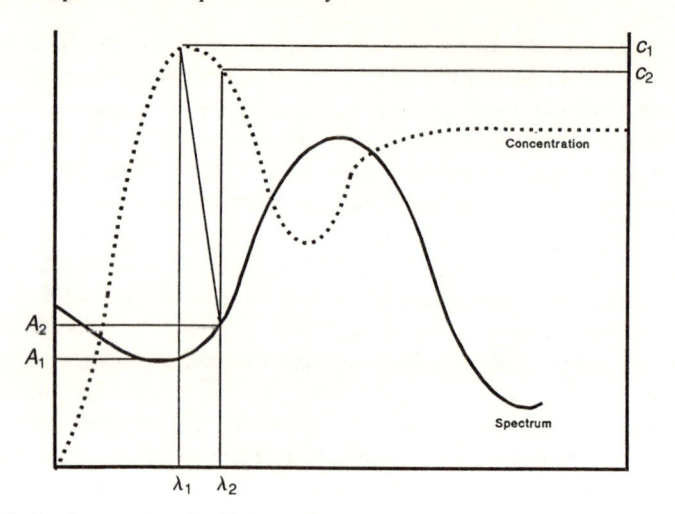

Fig. 4.3  *Mode of operation of a Kalman filter.*

The value of $Q$ can be altered during a calculation, and is achieved with an 'adaptive Kalman filter'. For example, during a calculation the sensitivity of the model to unexpected effects may need to be decreased. This may be in circumstances where two components are expected but a third is present contributing significantly in part of the spectrum. As the filter approaches this region then the differences between the actual and calculated absorbances will increase due to the third component. If $Q$ is increased then the filter will not force large changes in the calculated concentrations to compensate for the unexpected third component and a better estimate of the concentrations of the two expected components will be obtained. An additional benefit of this approach is that the residual spectrum calculated from the final concentrations is a good estimate of the unexpected component. At the time of writing only one instrument manufacturer is using the Kalman filter in a commercial spectrophotometer.

## 4.6  Multicomponent calibration

The techniques discussed above have used the spectra of pure components as the standards for the calculations. This presumes that in the sample mixture the spectra of the individual components are identical or at least very similar to those in the standards. In many cases and especially in dilute solutions this is the case. But when interaction between components does take place and this has an effect on the mixture spectrum the use of pure standards will not give reliable concentration estimates. Other calibration strategies and multicomponent data analysis techniques must be employed. [5]

If components are interacting in the mixtures to be assayed then these interactions must be introduced into the multicomponent calculations if they are to be successfully modelled and their effects minimized in the final calculation of the concentration estimates. This means that the assay must be calibrated using mixed standards. These mixed standards must be created or obtained such that they contain sufficient chemical, physical or other property information about the chemical system under investigation if a robust model is to be generated. In quantitative applications the calibration samples must represent the working range of concentrations of the components to be assayed. They must also be chosen to bring out any interaction effects between the components that may have spectroscopic effects in the concentration range of interest.

This calibration set of samples must contain all known sources of chemical, matrix or background variances likely to be seen in the test samples to be analysed.

Considerable thought and effort must be put into the preparation of these standards and the techniques of Experimental Design are employed to suggest a scheme for the standards. These techniques are widely available and represent an efficient way of maximizing the information present in a minimum number of calibration standards.

## 4.7 Principal component analysis and regression

Principal components analysis (PCA) is a data reduction technique aimed at generating an overview of the dominant information or variance patterns in a set of data [6]. These are the relationships between spectra and the relationships between wavelengths in the spectra. PCA sets out to express the main information in the calibration spectra in terms of a smaller number of variables. Factor analysis is used to reduce the number of dimensions in the calibration data set by extracting from them a set of factors or latent variables which represent the number of independently varying contributions to the original variance in the data set.

These factors are obtained from eigen analysis of the data covariance matrix and represent the underlying pattern of variance in an easily interpreted form.

The data covariance matrix $Z$ is

$$Z = AA' \qquad (15)$$

where $A$ is the absorbance data matrix, and $A'$ is the transpose of matrix $A$.

These factors are common to all the spectra. The amount, or score of each factor needed to reconstruct a spectrum varies from spectrum to spectrum. To model exactly the input data in terms of these factors and loadings requires as many factors as original spectra.

The power of PCA lies in its ability to discard those factors which contribute only to the noise and compress the data to several significant factors.

Eigenanalysis of the data 'ranks' the factors in order of decreasing eigenvalue, the first factors accounting for the most variation in the data, later factors accounting mainly for noise in the data. Techniques exist to identify the cut off point between significant and insignificant factors. Often complex data can be successfully modelled with 1 or 2 factors.

The original data matrix is therefore represented as

$$A = FS + E \qquad (16)$$

where $A$ is the matrix representing the absorbance values of the calibration spectra, $F$ is the 'factor' matrix, $S$ is the 'score' matrix, and $E$ is the unmodelled noise in the data.

Multiple linear regression is used to establish any correlations between the factor loadings and the composition of the calibration spectra and build a model of the mixture. This model, or correlation equation is then applied to the factor loadings of the test sample spectra to predict their composition.

## 4.8  Summary

For many applications UV–visible spectroscopy is capable of analysing the composition of both simple and complex mixtures without resort to a chromatographic front end, separation of the individual components being performed with mathematics rather than physical chemistry. The technique lends itself to the use of automated sample preparation and integration into a fully automated and controlled assay system. In this form UV–visible spectroscopy is assured of its place in both the modern laboratory and in the demanding applications of process control.

## References

1. Howell, J.A. and Hargis, L.G. (1990) *Anal. Chem.* **62**, 155R.
2. Martens, H. and Naes, T. (1989) *Multivariate Calibration*, John Wiley & Sons, New York.
3. Knowles, A. and Burgess, C. (1984) *Practical Absorption Spectrometry*, Chapman and Hall, London.
4. Tranter, R.L. (1990) *Anal. Proc.* **27**, 134.
5. McClure, G.L. (1987) *Computerized Quantitative Infrared Analysis*, ASTM Publication 934, Philadelphia, PA.
6. Malinowski, E.R. and Howery, D.G. (1980) *Factor Analysis in Chemistry*, John Wiley & Sons, New York.

# 5 UV–visible spectroscopic libraries

## 5.1  Introduction

The traditional role of UV–visible spectroscopy is that of quantitative analysis. Because of the potentially high molar absorbtivities of many chromophores the technique is particularly effective in detecting and measuring very low concentrations of species. It is not usually thought of as a detailed structural information technique because of the relatively featureless spectrum compared to that recorded in the mid-IR region. But qualitative information is present in UV–visible spectra and it can be put to good use in applications investigating the differences or similarities between spectra [1]. These applications involve setting up spectral databases or libraries and the use of mathematical means of comparing spectra.

Spectral libraries are collections or compilations of spectral data brought together for some purpose. They are generally, but not exclusively, single type spectra, that is all UV or NIR or mid-IR spectra. Mixed technique libraries do exist and are usually compiled from the spectra of complementary techniques such as mid-IR and mass spectrometry data. These data libraries are generally used as a reference source of spectra against which the spectrum of an unknown sample is searched using some comparison metric. Applications for spectral databases are now expanding swiftly with the increasing use of chemometric data analysis techniques.

UV–visible libraries are not yet widely established but some examples of their use are as follows:

1. Identification of an unknown material from its spectrum.
2. Confirmation that a spectrum belongs to a group of similar spectra.
3. Spectrometer control purposes.
4. Transfer of calibrations between instruments.

This chapter covers the setting up and creation of a database, spectral storage and a discussion of the types of algorithms required to compare spectra objectively.

## 5.2  Creating a spectroscopic database

Pivotal to the success of a spectroscopic database or library application is the validity of the spectral data used to create or calibrate it. Without quality spectra the data in the library is of small use and little confidence in the end results can be expected.

This validity is of two types: chemical validity and spectroscopic validity [2].

### 5.2.1  *Chemical validity*

Clearly the authenticity of the original sample materials used to produce the spectral data must be known. Without this the uses of the library are compromised even before any spectroscopy takes place.

### 5.2.2  *Material identity*

Materials should be from a reputable source. This may be from a vendor or from within your own organization. The purity or impurity profile of the material should be known or established before it is introduced as an entry in the library. Documents confirming identity and purity should be carefully preserved as part of the validation information for the library.

### 5.2.3  *Sample preparation*

Care must be given to choosing the appropriate sample preparation technique for the materials under investigation. Ideally this should be standardised for all the entries in the library but in a large library this may not be practicable. In most cases samples to be introduced into the library will be run as solutions. The choice of solvent is critical as solvent transmission range and solvent–solute interactions can greatly modify the spectrum. When deciding what is the appropriate solvent for the exercise it is essential that solvent composition, pH, ionic concentration are identical between samples if meaningful comparisons are to be made.

## 5.3  **Spectroscopic validity**

It is important that the measurements made on the sample after its preparation and presentation to the spectrophotometer are valid. The instrument must be performing within its nominal specification. To ensure this a rigourous testing of the instrument must be performed prior to any series of measurements. This is standard practise in many laboratories and without it both the short term and long term usefulness of the spectroscopic data must be questioned.

### 5.3.1  *Wavelength calibration*

Wavelength standards represent the most practical method of measuring the wavelength accuracy and precision of an instrument. Holmium and didymium oxide glasses are used extensively for the routine testing of low resolution spectrophotometers. An aqueous solution of holmium oxide in perchloric acid offers a series of narrow absorption bands for calibrating higher resolution instruments. The most accurate calibration method is to introduce the light from a discharge lamp into the optical path of the spectrophotometer. Precise values for the location of emission lines are available from standard reference works. For short wavelengths mercury lines are useful (253.65 and 184.96 nm), potassium (770.0 and 766.5 nm) for mid range wavelengths and lithium (670.8 and 610.4 nm) for long wavelengths. As many spectrophotometers use a deuterium lamp as a light source they contain a ready source of calibration lines at 486.00 and 656.10 nm.

### 5.3.2  *Absorbance calibration*

For routine calibration of the absorbance measurements of a spectrophotometer a vast array of standards are available and the user should choose those that are convenient for everyday use. Glass absorbance standards made from Schott NG-4 glass are now established as a reliable source of known absorbance values ranging from approximately $1.9A$ to $0.1A$. For absorbance measurements down in the UV-region glasses are not transparent and are replaced by evaporated metal on quartz or fused silica substrates. Liquid absorbance standards, although requiring preparation represent a very useful means of checking a spectrophotometer and better represent the normal samples that an instrument will measure than the solid samples mentioned above. They do however require very careful preparation and handling and may be limited in their applications.

### 5.3.3  *Stray light*

A common source of absorbance measurement error in spectrophotometers is that of stray light. This is the emergence of radiation from the monochromator of wavelengths widely differing from that indicated by its bandpass. This has the effect of producing an apparent absorbance error in the readings and it must be identified if measurements are to be made reliably. Identifying stray light is much easier than quantifying it as the processes responsible for stray light are very much sample dependent. But routine measurement of stray light is easily performed with a cut-off filter.

## 5.4  **Data storage**

The library of spectra must be stored in some format on the computer system of choice. In many cases the spectroscopist will be using a proprietary data system and software package to generate the library and no choice of storage format will be available; the manufacturer of the equipment will impose its own. This is acceptable if the transfer of spectral data between instruments or systems and the size and use of the data base is not compromised. If spectra need to be transferred between different manufacturer's systems specific data formats are a limitation. The format in which the digital spectroscopic data are stored in the library must be carefully chosen to optimise the use of the library. To make a library 'future proof' full range raw spectral data should be stored, allowing the searching algorithms invoked by the library software to be changed to improve the application without having to re-run samples.

### 5.4.1 *The JCAMP-DX transfer protocol*

Transferring spectra between instrument systems or between software packages within a computer requires a common file format for data interchange. To this end a common standard data format has been established and is being promoted as an 'industry standard'.[3] This is the JCAMP-DX file format. JCAMP is an acronym for the Joint Committee of Atomic, Molecular and Physical Data. It represents the first official attempt to realise some form of standardisation within the field of spectroscopic data and is now adopted by IUPAC. Primarily aimed initially at the FTIR users its protocols are easily used with UV-visible data.

The main criteria behind the JCAMP-DX specification are:

1. representation of the digital data without loss of precision;
2. inclusion of detailed sample information;
3. internal documentation readable by humans;
4. acceptable by a wide range of computer systems; and
5. expandable to meet future needs.

The JCAMP-DX file is a sequential ASCII text file which consists of a series of 'labelled data records' (LDR). Each LDR can occupy as many lines as necessary to represent the data. Each line can only contain a maximum of 80 characters. The LDR is composed of a data label and its associated data set, and a series of these labels is assembled in order to give the JCAMP-DX file. An example is shown in Fig. 5.1.

In summary the quality in both chemical and spectroscopic terms of the spectral data in a library is all-important. Producing a high quality spectroscopic library is expensive in both time and money but if done carefully should be a long-lasting resource.

## 5.5   Techniques for searching spectral libraries

Classical spectral libraries exist in hardcopy paper form bound into books. Traditionally the spectroscopist obtains the spectrum of the material under investigation, plots it onto a sheet of chartpaper and compares it with those in the book. This matching is usually by eye and is highly subjective. To be accomplished at it normally requires years of experience. Much more rigorous and objective approaches can be taken when a library exists in a digital form. Spectra can be compared mathematically for their similarity or differences using a variety of numerical techniques. Here the principle is usually to produce a 'hit list' of spectra from the library which best match the test spectrum.

```
'##TITLE = Methyl Orange at pH 5.2 temperature 15C
##JCAMP-DX = 4.24
##DATA TYPE = UV/VIS
##ORIGIN = ACME Laboratories
##XUNITS = Nm
##YUNITS = absorbance
##XFACTOR = 1
##YFACTOR = .00001
##FIRSTX = 190
##LASTX = 510
##NPOINTS = 321
##FIRSTY = .24178
##XYDATA = (X++(Y..Y))
190 24178 −9036 3209 29134 7373 16356 −7539 −8641
14157 49001 −27074 −28615
202 −28822 25461 1576 8772 −12463 7545 1978 10783
−13365 −27379 9592 −18320
   .
   .
   .
   .
   .
498 39755 39146 38368 37257 36638 35570 34700 34015
33107 32256 31432 30257
510 29477
##END =
```

Fig. 5.1  *Example of a JCAMP-DX file.*

This 'hit list' is usually generated from evaluating what is known as a similarity metric. This is an algorithm which compares the unknown spectrum against candidates in the library and evaluates their similarities in terms of a number. For identical spectra this number will usually be zero. As the unknown and candidate spectra become increasingly different in their profiles then the evaluated number will rise. The essence of the approach is deciding which of the many evaluation metrics is optimum for the application and deciding just when two spectra are 'the same'. In the 'real world' we can only ever expect non zero evaluations of the similarity metric as two spectra will never be absolutely identical, if only due to instrumental noise.

### 5.5.1   *Data normalization*

Before comparison algorithms can be discussed it is important that the topic of spectral data normalization is understood. For any library search application to be successful it is important that 'like' is searched against 'like'. If comparison algorithms are to be used efficiently then in many cases the spectral data must undergo a normalization process before they are evaluated for similarity. This removes or minimizes sources of variance between spectra which are not of interest in the application. An example of this is the effect of sample preparation variances in making up solution spectra. If solution spectra of materials are to be compared against those in a library it is important that the spectroscopic effects due to concentration differences between spectra in the library and those in the unknown spectra are removed if the comparison metric is to measure chemical similarities or differences effectively. Effects due to noise on the spectra can also distort comparison metrics.

Mathematical techniques have been developed to pre-treat spectra to minimize many types of unwanted or unneeded variances [4]. Several of these are listed below, using the standard notation:

$X_j$ is the absorbance of spectrum $X$ at wavelength $j$.
$X_j^*$ is the absorbance of the normalised spectrum $X^*$ at wavelength $j$.
$Max_X$ and $Min_X$ are the maximum and minimum absorbance values, respectively, of spectrum $X$.
$m_j$ and $s_j$ are the mean and standard deviation values, respectively, of a group of spectra $Y$ at wavelength $j$, if available.

## 5.6 Spectral scaling

A spectrum can be scaled relative to a variety of spectral features. These may be internal features such as the maximum and minimum intensity of the spectrum itself or external such as features from a reference spectrum or group of spectra. The technique of choice will depend on the source of the variance needed to be minimized for that particular application.

### 5.6.1 Maximum scaling

This normalizes a spectrum with respect to its internal maximum absorbance value. This is useful in scaling data subject to Beer–Lambert law effects such as variable concentration or pathlength.

$$X_j^* - X_j / Max_X$$

### 5.6.2 Range scaling

This normalizes data as above but removes baseline offset effects.

$$X_j^* = \frac{(X_j - Min_X)}{(Max_X - Min_X)}$$

### 5.6.3 *Autoscaling*

If mean and standard deviation values are available from some reference group of spectra then they can be used to scale a spectrum to unit variance. These type of values are readily obtained for a spectrum if a diode array spectrophotometer is used to collect the data. Failing this, replicate spectra of a sample can be measured on a conventional spectrophotometer and the values calculated.

For external referencing the mean and standard deviation can be calculated from a reference set of spectra and applied to the sample spectrum to be normalized.

Autoscaling is a powerful technique for pre-treating data if the effects of shape differences between spectra are important. Absolute intensity differences between spectra are minimized by subtracting the mean, and differences in noise between spectra are minimized by dividing by the standard deviation.

$$X_j^* = \frac{(X_j - m_j)}{s_j}$$

## 5.7 Comparison techniques

### 5.7.1 *Feature selection: peak matching*

The simplest form of computerised library search is to identify unique features in the unknown spectrum and match them against those belonging to the spectra in the library, the best match being the most likely candidate. These features are generally the wavelength position or the position and intensity of the major absorption features in the spectra. This selection of features greatly reduces the amount of data that has to be processed by the matching algorithms but has the drawback of sacrificing what may be a considerable amount of unique spectral information, resident in peak shapes and baseline slopes. Peak position and intensity information is a very 'user friendly' form of spectral information as the data can be readily understood and interpreted by the chemist, often in terms of functional group or chromophore chemistry [5].

The comparison of the spectral features of an unknown spectrum and a library candidate spectrum can be performed in several ways. An *absolute match* scoring is intended for applications where the spectrum is presumed to be a single entity and not a mixture. Here a positive weighting is given to features present in both the unknown and candidate spectrum, and a negative weighting is applied if absorption features are absent in either the library or unknown spectrum. In the case of mixture spectra or spectra known to include impurities then a *positive match* scheme is often more useful. Here only features present in both the unknown and candidate spectra are counted in the evaluation.

70   UV–visible spectroscopic libraries

The feature selection algorithms used to generate the peak/intensity data for these comparison routines must be able to accurately determine the presence of a peak above the noise level in the spectrum. This is not a real problem with UV–visible spectra but can be critical in mid-IR applications where important weak features may have an intensity comparable to the noise in the spectrum.

### 5.7.2   Feature selection: Mahalanobis distances

Another feature selection technique uses the spectral data in the library entries to generate a set of multidimensional axes into which the library spectra are projected as a series of points. These axes are absorbances at wavelengths which represent the optimum differences between the spectra in the library. Once projected into this hyperspace the multidimensional distances between the points representing the spectra are then a direct measure of the difference between the spectra. The closer the points are the more similar the spectra are at the wavelengths used as the axes. An unknown spectrum is then projected into this hyperspace and the multidimensional distance between it and the library entries evaluated to identify the most likely candidate spectrum.

One example of this feature selection approach to spectral plotting is the use of 'Mahalanobis distances' as a comparison metric.

The spectra shown in Fig. 5.2 represent a typical library of reference materials. The data can be greatly simplified by replotting. If three wavelengths are selected to form the axes of a three dimensional plot and the corresponding absorbance values of each library spectrum are plotted in these axes we see that the spectra plot as a set of discreet points unlike the overlapped spectra. This can be seen in Fig. 5.3. Different spectra of the same material will plot as a cluster of points due to spectral noise. This noise may be instrumental or result from sample preparation

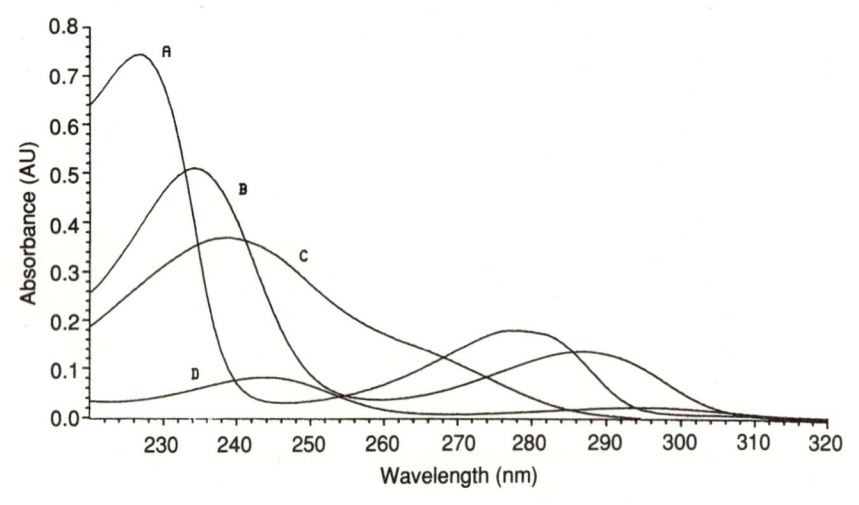

Fig. 5.2   *Spectra of a typical library of reference materials.*

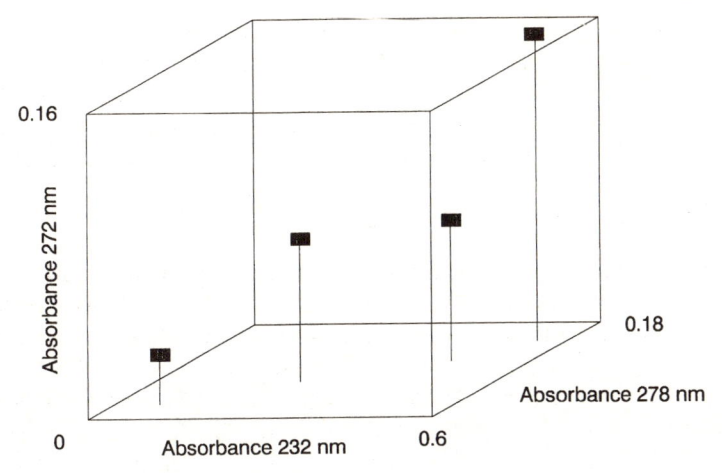

Fig. 5.3 *Plot of absorbance at 278nm against 232nm against 272nm for a UV spectral library.*

or presentation variances between samples but it represents a source of variance that can be used to calculate statistical confidence intervals around a group of points in terms of standard deviations. The standard deviation contour around such a variant group as this is called the Mahalanobis unit and is used as a direction dependent metric in evaluating spectral differences. This can be used to obtain the probability, in terms of standard deviations, of an unknown spectrum belonging to that group of spectra. The Mahalanobis distances between the points or clusters of points represent a measure of their spectral differences.

The wavelengths used in this plot are chosen or calculated to maximize the distances between different materials in the library. Often two wavelength axes will not be sufficient to separate materials adequately in terms of the number of standard deviations (Mahalanobis units) they are apart and others must be added. Although the complexity of the calculations rises rapidly as the number of dimensions increases a computer has little problem in handling the exercise. In many practical uses of this approach up to 15 dimensions must be used to plot the data in before different groups of spectra are adequately resolved from each other.

Once a library is 'calibrated' in such a way it can then be used to identify unknown spectra. The spectrum of the unknown is obtained and plotted in the library space of wavelengths described above. The multidimensional Mahalanobis distances between the unknown and the library entries are then evaluated, the unknown material being assigned to a group with a distance less than some cut off value determined from the standard deviation contours of the groups in the libraries.

The vector distance of an unknown spectrum from the centroid of a group in the library is given by:

$$D_k^2 = (X - X_k) M_i (X - X_k)'$$

where

$D_k$ is the Mahalanobis distance of a test sample $X$ from the centroid of the library group $X_i$.
$X$ is the multidimensional position of the unknown spectrum in the wavelength space of the library.
$X_k$ is the multidimensional position of the library group $k$ in the wavelength space of the library.
$(X - X_k)'$ is the transpose of the vector $(X - X_k)$
$M_i$ is the inverse of the library spectra variance — covariance matrix and defines the distance measures for the Mahalanobis metric calculations.

The wavelengths used to define the multidimensional space in which these calculations are performed can be selected by evaluating $(1/D_{kl})$ for a minimum.

$D_{kl}$ is the Mahalanobis distance between the groups $k$ and $l$ in the library. The greater the magnitude of the intergroup distances the more discriminating power the library will have between the entries. This is especially important in case where the library may consist of very similar spectra.

The Mahalanobis distance approach only uses what may be a very small amount of the total spectral data, perhaps only 10 wavelengths out of 350 available, and it is conceivable that important spectral differences between materials could contribute to their spectra at wavelengths not evaluated in the Mahalanobis distance calculations. It nevertheless represents an extremely rapid, robust and statistically sound means of searching large libraries.

### 5.7.3 Full spectral matching

The most incisive means of discriminating between spectra are those which use all the available spectral data. These are generally mathematically intense in nature when applied to large libraries.

Here the spectrum of an unknown material is compared point by point with those in the library using a 'full spectral matching algorithm'.

## 5.8 Comparison algorithms

Comparison techniques using all the spectral information usually employ some sort of distance metric which can simplify differences at many wavelengths into a single number for evaluation or ranking purposes [5]. Some of the metrics are listed below:

$U_j$ is the absorbance of unknown spectrum $U$ at wavelength $j$.
$X_j$ is the absorbance of the candidate spectrum $X$ at wavelength $j$.

### 5.8.1 Euclidian and Manhattan metrics

Two of the simplest and most readily understood metrics are the Euclidian and Manhattan. These metrics use simple geometry to evaluate the 'distance' between

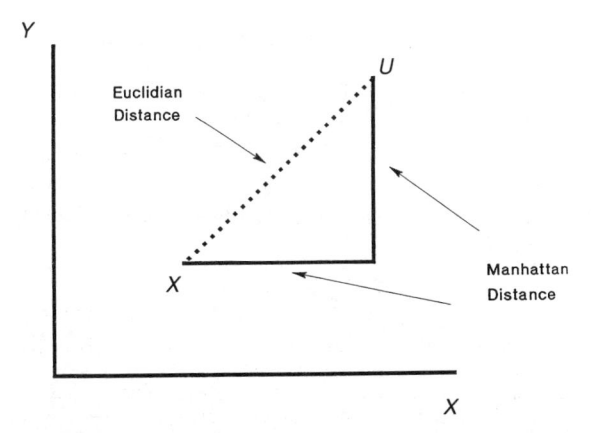

Fig. 5.4  *Distance metrics. Euclidean: $\{\Sigma(X - U)^2\}^{1/2}$; Manhattan: $\{\Sigma X - U\}$.*

two spectra. The 'shorter' the distance the better the match between the spectra, the 'larger' the distance the poorer the match. A perfect match would be 0 'distance'. The simplicity of these metrics is illustrated in Fig. 5.4 and specified below:

$$d_{UX} = \left(\sum (U_j - X_j)^2\right)^{1/2} \quad \text{(Euclidian metric)}$$

$$d_{UX} = \sum |U_j - X_j| \quad \text{(Manhattan metric)}$$

### 5.8.2  Spectral correlation

The comparing of spectra can be performed using linear regression techniques [6]. If the unknown spectrum is treated as a set of dependent variables and the candidate spectrum as the set of independent variables it is possible to perform a linear regression of the two. This yields important spectroscopic information.

The correlation coefficient, $r$, or $r^2$, is a measure of the spectral similarity in terms of peak positions and relative intensities.

The slope, $m$, of the regression line is a measure of the match of the baseline corrected absorption intensities, and the intercept, $c$, a measure of any baseline offsets between the spectra.

For a high quality match of spectra the correlation coefficient should be near to 1, the slope 1 and the intercept 0. A slope significantly different from unity indicates differences in spectral features, and extra or missing peaks. A good correlation with a non-unity slope indicates a concentration or Beer–Lambert Law difference between the spectra.

The intercept, $c$, is a measure of the difference in baselines between the spectra being compared. Baseline offsets due to scattering or cell effects will manifest themselves as nonzero values of $c$.

## 5.9 Comparison metrics applied to a demonstration library

A small UV spectral library is shown in Fig. 5.2. This contains four different solution spectra A, B, C, and D collected on a diode array spectrophotometer. Fig. 5.5 shows a spectrum ($U_1$) that will be used as an 'unknown' to test the metrics described above. For this example the 'unknown' spectrum is another sample of C. Each spectrum consists of 101 data points collected over the range 220 to 320 nm.

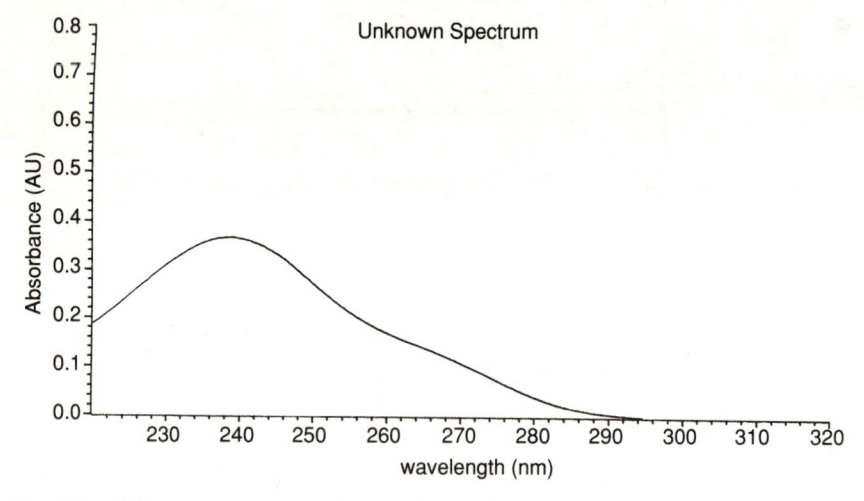

Fig. 5.5 Unknown spectrum used to test the metrics.

An integral part of the validation of the spectral library and its search algorithm and metric is the evaluation of the intergroup distances in the library. This will indicate the efficiency of the metric in distinguishing between the different spectra in the library. If a metric cannot reliably discriminate between the spectra in the library it cannot be expected to give unambiguous results when evaluating an unknown spectra against the library.

The interlibrary metrics in Tables 5.1 and 5.2 show that the spectra in the library are easily discriminated between using the three metrics described above.

The unknown spectrum $U_1$ (Tables 5.3 and 5.4) is now evaluated against the library using the comparison of three metrics.

Clearly all three metrics can identify the unknown $U_1$ as belonging most likely to group C. The Manhattan and Euclidian distances evaluate with very low values for the group C to $U_1$ match, much lower than the next closest group B. The correlation metric evaluates the match between C and $U_1$ as perfect, when rounded up to four significant figures. This is an ideal situation for these metrics as the only differences between the spectra are those of a chemical nature, the preparation of the sample and the running of the spectrum being tightly controlled. In the real world, and particularly with UV spectra, differences

**Table 5.1**  *Manhattan and Euclidian metrics*

Interlibrary distances

| From group | To group | Manhattan | Euclidian |
|---|---|---|---|
| A | A | 0.00 | 0.00 |
| A | B | 23.01 | 3.05 |
| A | C | 30.75 | 3.59 |
| A | D | 30.50 | 4.17 |
| B | B | 0.00 | 0.00 |
| B | C | 15.18 | 1.37 |
| B | D | 16.31 | 2.05 |
| C | C | 0.00 | 0.00 |
| C | D | 17.74 | 1.79 |
| D | D | 0.00 | 0.00 |

**Table 5.2**  *Spectral correlation regression*

Interlibrary correlation/regression

| From Group | To Group | $r^2$ | m | c |
|---|---|---|---|---|
| A | A | 1.000 | 1.000 | 0.000 |
| A | B | 0.709 | 0.373 | 0.025 |
| A | C | 0.527 | 1.091 | 0.046 |
| A | D | 0.556 | 5.793 | 0.026 |
| B | B | 1.000 | 1.000 | 0.000 |
| B | C | 0.788 | 0.858 | 0.020 |
| B | D | 0.803 | 4.340 | 0.006 |
| C | C | 1.000 | 1.000 | 0.000 |
| C | D | 0.815 | 0.162 | 0.005 |
| D | D | 1.000 | 1.000 | 0.000 |

**Table 5.3**  *Unknown (U1) — Library distances*

| From Group | Manhattan value | Euclidian value |
|---|---|---|
| A | 30.63 | 3.59 |
| B | 15.06 | 3.59 |
| C | 0.15 | 0.01 |
| D | 17.68 | 1.79 |

**Table 5.4**  *Unknown (U1) — Library correlation/regression*

| Group | r | m | c |
|---|---|---|---|
| A | 0.527 | 0.254 | 0.024 |
| B | 0.788 | 0.724 | 0.004 |
| C | 1.000 | 1.000 | 0.000 |
| D | 0.815 | 4.10 | −0.003 |

## 76 UV–visible spectroscopic libraries

between spectra may be from sources other than chemical. Typically they arise from poorly controlled sample preparation and bad spectroscopy. The effects of these form of variances can easily be shown.

The effects of concentration differences and baseline offsets on the evaluation of an unknown spectrum against a library can be seen if the spectrum $U_2$, shown in Fig. 5.6, is searched against the library. This spectrum is of material C but has a low concentration relative to that of spectrum $U_1$; $[U_2] = 0.3[U_1]$. A baseline offset of $0.050A$ has also been added. This is similar to that seen from scattering effects or cell placement errors.

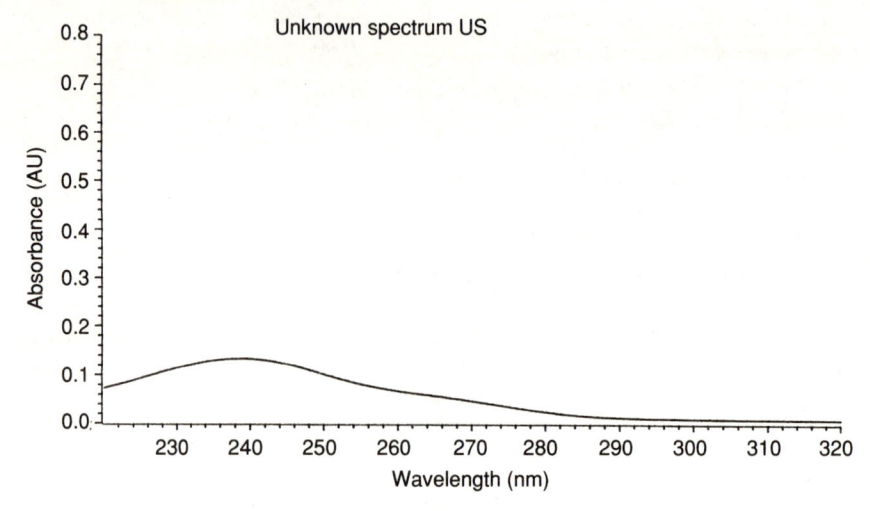

Fig. 5.6  *Low concentration with unknown spectrum.*

The evaluation of the Manhattan and Euclidian metrics, Table 5.5, now suggests group D as the most likely candidate for $U_2$ using the nearest neighbour approach of classification. This is clearly wrong but the simple metrics cannot distinguish between spectral pattern and absolute differences.

The results of the 'correlation metric', however, predict the most likely match still being group C with $U_2$, Table 5.6. The correlation coefficient is evaluated as 1.000, rounded up to four significant figures, but the slope $m$ is now 0.300

**Table 5.5**  *Unknown (U2) — Library distances*

| From Group | Manhattan value | Euclidian value |
|---|---|---|
| A | 36.94 | 3.85 |
| B | 22.41 | 1.72 |
| C | 21.92 | 1.41 |
| D | 16.77 | 1.00 |

**Table 5.6** *Unknown (U2) — Library correlation/regression*

| Group | r | m | c |
|---|---|---|---|
| A | 0.527 | 0.076 | 0.057 |
| B | 0.788 | 0.217 | 0.051 |
| C | 1.000 | 0.300 | 0.050 |
| D | 0.815 | 1.230 | 0.049 |

instead of 1.000 for $U_1$, and the intercept $c$ is 0.050 instead of 0.000 for $U_1$. The slope is a measure of the concentration difference between the two spectra and the intercept a measure of any offset between the two spectra.

Spectral matching using very large libraries can be slow, as spectra are essentially matched spectrum by spectrum and point by point. Time savings can be achieved by restricting the wavelength range of the comparison, lowering the data point resolution of the digitised spectra or reducing the spectral information by use of compression techniques such as Fourier transforms [7].

Many spectral matching algorithms exist, and the choice of which to use is highly application-dependent. In the cases above the correlation metric copes best when matching spectra with concentration and offsets differences, the Manhattan and Euclidian metrics being best when dealing with absolute matches, i.e. dealing with the question 'Are these spectra identical, similar or dissimilar?'

# References

1 Brown, C.W. and Donahue, S.M. (1988) *Applied Spectroscopy*, **41**, 7.
2 Knowles, A. and Burgess, C. (1984) *Practical Absorption Spectrometry*, Chapman and Hall, London.
3 McDonald, R.S. and Wilks, Jr, P.A. (1988) *Applied Spectroscopy*, **42**, 1.
4 Todeschini, R. (1989) *Chemometrics and Intelligent Laboratory Systems*, **6**.
5 Whitfield, R.G., Gerger, M.E. and Sharp, R.L. (1987) *Applied Spectroscopy*, **41**, 7.
6 Salmin, P.A., Cornelis, Y. and Bartels, H. (1988) *Chemometrics and Intelligent Laboratory Systems*, **3**.
7 Kawata, S., Noda, T. and Minami, S. (1987) *Applied Spectroscopy*, **41**, 7.

# 6 Spectra-structure correlation

## 6.1 Origin of spectra

The value of electronic spectra in the determination of structure is generally underestimated today, largely because much of the detailed knowledge of spectra-structure correlation and an appreciation of its applicability has been forgotten, except by a few practitioners. Fortunately, a substantial portion of that knowledge has been collated and discussed somewhere. It is hoped that the list of selected texts (Section 6.4) will provide an entry into the vast literature.

At the very minimum, it is useful to be able to understand the relation between the structure of a substance and its electronic spectrum and so avoid wasting time trying to detect the undetectable, for example in high performance liquid chromatography (HPLC). It is better still to be able to judge whether a given spectrum is compatible with an assumed structure.

For a substance to have an observable spectrum it must have an excited state as shown in Fig. 6.1 with an energy difference from the ground state corresponding to an accessible wavelength of the electromagnetic spectrum:

$$\Delta E = h\nu \qquad \bar{\nu} = \frac{1}{\lambda} \qquad \nu = \frac{c}{\lambda}$$

where $\Delta E$ is the energy difference, $\nu$ the frequency, $\lambda$ the wavelength, $\bar{\nu}$ the wavenumber, and $c$ the velocity of light.

|  | Vacuum UV |  |  | Visible |
|---|---|---|---|---|
|  | 12.4 | 6.2 | 3.1 | 1.5 eV |
| $\lambda$ | 100 nm | 200 nm | 400 nm | 800 nm |
| $\bar{\nu}$ | 100 000 cm$^{-1}$ | 50 000 cm$^{-1}$ | 25 000 cm$^{-1}$ | 12 500 cm$^{-1}$ |

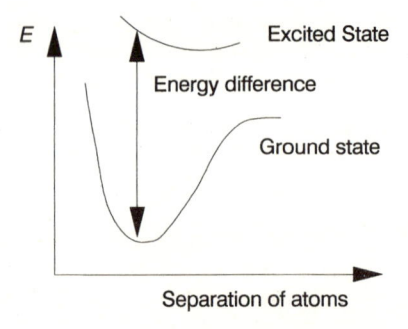

Fig. 6.1 *Relationship between energy and wavelength/wavenumber in the electromagnetic spectrum.*

## 6.2 Common absorption bands

### 6.2.1 Inorganic compounds

Transition metal ions have such excited state levels: from their colours we know that they must have absorption bands in the region 400–700 nm. Rare earths in particular (holmium perchlorate solutions; Nd–YAG crystals) have their longest wavelength transitions in the near infrared region. The ions of other heavy elements have transitions in the ultraviolet region, but alkali metal ions and many of the simplest compounds of the first row elements (e.g. $H_2O$, $NH_4^+$, $CH_4$, $ClO_4^-$) are essentially transparent in the regions accessible to most UV spectrometers.

### 6.2.2 Aliphatic compounds and their use as solvents

The pure saturated hydrocarbons, alcohols, ethers and amines have their longest wavelength peak well below 200 nm, and are therefore transparent above 200 nm. Chlorinated hydrocarbons also have transition maxima below 200 nm, but the tail of the absorption of the neat substance cuts off radiation below 240 nm at the usual pathlengths. The concentration difference between solute and solvent, a factor of about 100 000, always needs to be remembered. Useful solvents are therefore limited to those mentioned and aqueous solutions containing transparent buffers. However, because UV absorbing species are usually detectable at concentrations below 10 mg dm$^{-3}$ and sometimes several orders of magnitude below that, the concentration of an adequate buffer system often does not need to be very high.

### 6.2.3 Chromophores

Most of the absorption displayed by organic substances[*] depends on the presence of double bonds. Such an absorbing system is called a chromophore. Ethylene absorbs about 165 nm, but substituting it with alkyl groups moves the absorption towards 200 nm. Such groups which cause a bathochromic shift, that is a shift to the red, are called auxochromes in dyestuff terminology. Alkyl groups are weakly auxochromic in this case.

---

[*] We often speak of 'the absorption spectrum of a molecule', but this statement contains the three great illusions of the usual practice of electronic absorption spectroscopy. Firstly, spectrometers do not measure absorption, but only transmission of radiation; secondly, they do not measure the spectrum in the sense that the *detector* does not know the wavelength, or even whether we remembered to put the sample in; and thirdly, that even if we loosely regard this as the absorption spectrum, it is the spectrum of the *species and its environment* that we measure. All molecules are sensitive to some extent to the solvent and to temperature and concentration, some extraordinarily so. But it is the pH in aqueous solution which can have a dramatic effect on many substances. Therefore solvent and conditions need to be taken into account when comparing spectra and collating substituent effects. On the other hand, spectral changes with pH, solvent, concentration and temperature can be put to positive use in spectral interpretation.

## 80 Spectra-structure correlation

Since the alkyl hydrocarbons are themselves transparent this increase in absorption wavelength must be due to the interaction of the alkyl group with the unsaturated system. In practice this results in small shifts which as in the case of substituted ethylenes are usually bathochromic but can be either negative and positive in other cases, ranging from about −5 nm to perhaps +7 nm per group, dependent on the system.

|  |  |  |
|---|---|---|
| $CH_3$—CH=CH—H | $CH_3$—CH=CH—$CH_3$ | $(CH_3)_2C=C(CH_3)_2$ |
| 177 nm | 179 nm | 196 nm |

Conjugating the double bond with further double bonds leads to increases both in intensity and wavelength of the absorption band in a systematic way, as is shown in Fig. 6.2. The increase in wavelength with increased chromophore lengths eventually levels off, although the limited stability of the higher polyenes generally prevents a complete investigation.

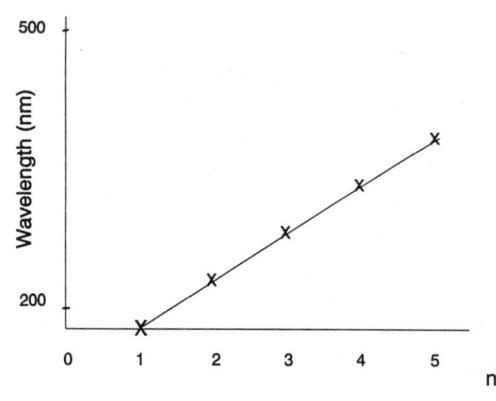

Fig. 6.2 *Relationship between wavelength of the absorption maximum and number of double bonds for conjugated molecules of the series $CH_3$—$(CH=CH)_n$ $CH_3$.*

The carotenoids, the yellow and red pigments of tomatoes, oranges, dandelions and most other yellow or orange flowers are conjugated systems with a particular alkyl substitution pattern shown in Fig. 6.3, and possibly with hydroxy, carbonyl and carboxyl substitution[4].

The carotenoids as well as the simpler linear systems have been systematically investigated so that their absorption wavelengths can be forecast by considering the incremental effects of the substituents on the basic structures.

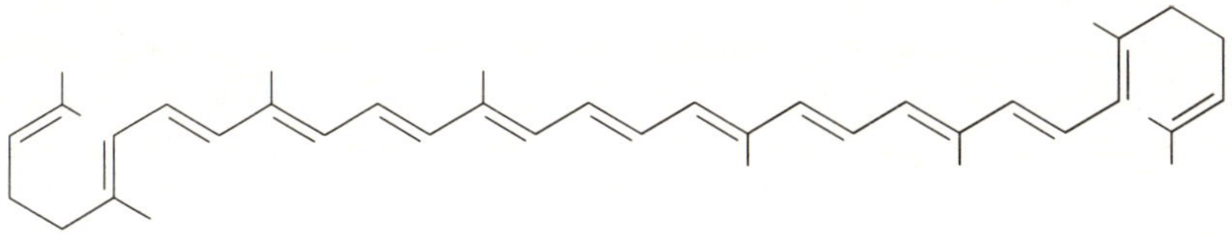

Fig. 6.3  *Lycopene*.

With carbonyl or carboxyl substitution it may be more reliable to consider the compound as a separate conjugated ketonic or acidic parent system and build up the alkyl or hydroxyl increments due to substitution on that parent system rather than to consider the more perturbing oxygen functions as substituents. The rules for the shorter chromophores are known as Woodward's rules[2], those for carotenoids, Zechmeister's rules[1], and those for steroids, Fieser's rules[3]. Although these rules are often discussed quite thoroughly in degree texts of organic chemistry, this is commonly the limit of what is presented about ultraviolet structure-spectral correlation. The calculated absorption maxima for such systems agree well with the observed ones, and are capable of considerable refinement in the case of well defined systems of related compounds provided substantial* steric effects do not interfere[4]. The absorption depends on the continuity, that is, the planarity of the conjugated system. Any substitution which interferes with this will reduce the intensity of the absorption dramatically and will generally, but not universally, reduce the absorption wavelength slightly[5].

If a charge separated system is now added to the conjugated system, an increase in wavelength occurs, for example Fig. 6.4. These are the cyanine dyes (usually with heterocyclic end groups) which absorb in the far red and are used to increase the sensitivity of photographic emulsions in the red or in the infrared[6].

There is continuing work directed to the production of dyes for the near infrared region in connection with military interests, and in electronics and in security printing as well as in photography[7]. However, a problem with the

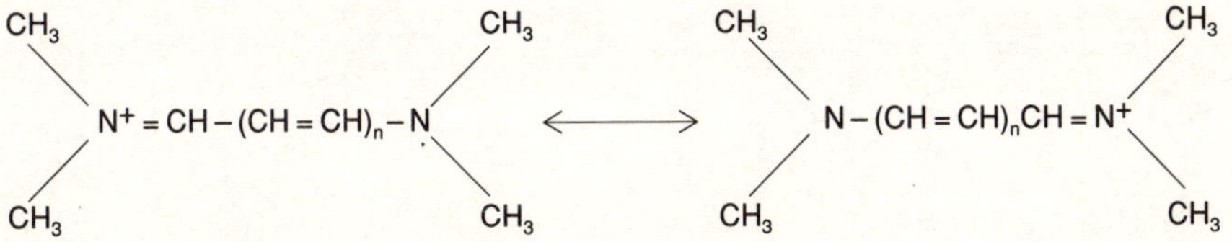

Fig. 6.4  *Resonance in charged conjugated structures*.

---

* 'substantial' because it has become clear, in recent years, that these rules are themselves due to a combination of steric and electronic effects.

longer linear conjugated systems is their instability. Turning the conjugated system into an aromatic system shortens the wavelength, but increases the stability dramatically. If the aromatic rings are replaced by heterocycles, with charge separations or with polarisable atoms, particularly sulphur, the necessary stability combined with long wavelength absorption can be achieved.

However, it is much more difficult to forecast the wavelengths of systems containing polarizable atoms than it is for those with only C, H, and O. In principle, the UV–visible absorption spectral maxima of any organic system containing only C, H, and O should be fairly closely estimable by sets of rules involving incremental changes from reference compounds. The flavonoid pigments (Fig. 6.5) which are responsible for the pink, purple and blue colours of most flowers[8, 9] and of the purple of beetroot are a particularly good example of this. The addition of alkali, or of certain coordinating ligands enables the pigment to be identified from the resulting spectral changes[8].

Molecular orbital/computer methods of calculating electronic spectra are being increasingly applied to more complex molecules, but the calculations are laborious and complex for any but the simplest molecules, such as ethylene and butadiene which are discussed in the standard texts. All the methods referred to below, although considered old-fashioned by mathematical spectroscopists and quantum chemists, have the advantage that they can be understood without any mathematics and although some aspects may be conceptually complex they relate spectra and structure in a 'chemically' satisfying way.

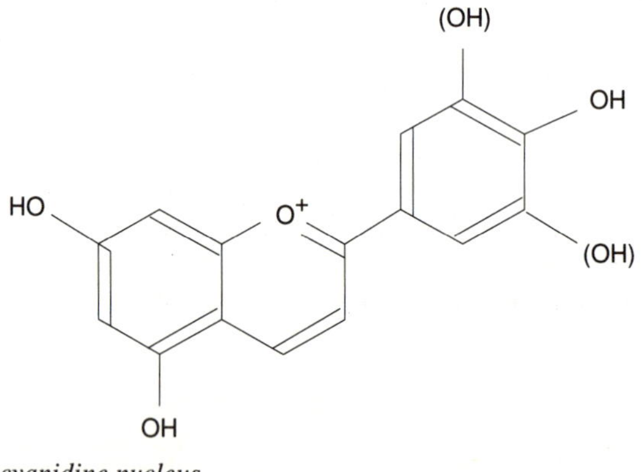

Fig. 6.5 Anthrocyanidine nucleus.

## 6.2.4 Aromatic and polyaromatic compounds

Turning attention now to benzene and the aromatic hydrocarbons, the spectra of the unsubstituted polycyclic aromatics can be calculated using 'electron in-a-ring' or free-electron theory methods[1, 10], although cross-conjugated systems cause problems. The position of the band maxima in these polycyclic aromatics is related to the length of the path of the conjugated system. Benzene itself has three absorption bands, that of the longest wavelength being, if the fine structure is wiped out, at 254 nm (Fig. 6.6).

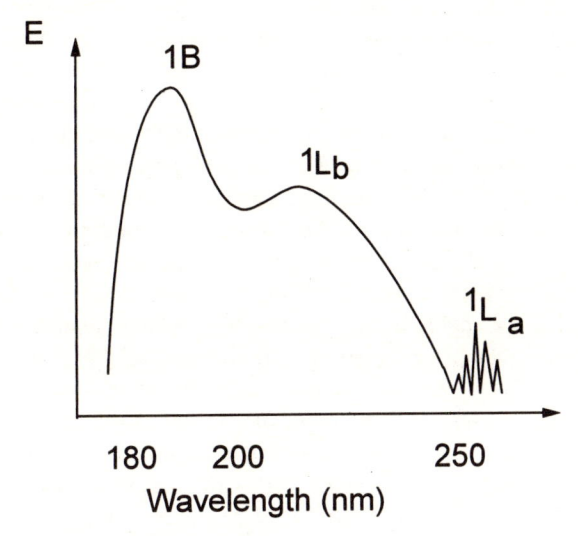

Fig. 6.6  *Schematic spectrum of benzene.*

It is weak because it is nominally a symmetry forbidden transition. It becomes more allowed upon substitution. Comparison of the wavelengths of the simple substituted benzenes allows a series of increments to be assembled[11] which can be transferred to other systems, for example, to polycyclic hydrocarbons or heterocycles, provided that no tautomeric, charge transfer or steric effects intervene[12]. Furthermore, with the same limitations, the effect of two or more substituents can be added vectorially, i.e. making allowance for the direction of the dipole, with a very close agreement between calculated and observed values, both for wavelength and intensity[13]. (Fig. 6.7).

$R = H, N^+R_3$
$CH_3$
$OH$
$NH_2, O^-$
$CHO$
increasing wavelength shift.

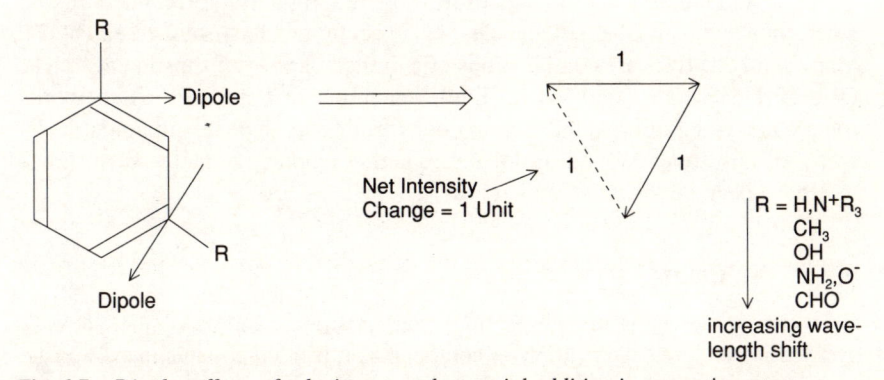

Fig. 6.7  *Dipolar effects of substituents and vectorial addition in aromatic systems.*

Note particularly the electron transparency of an $^{\pm}NR_3$ group: protonating an amine, provided the charge really stays on the nitrogen atom, causing the spectrum to revert to that of the parent.

The effects of steric hindrance on the electronic spectra have been systematically investigated in the case of diphenyls[14] but the principles are extendable to other systems. The UV spectra of tautomeric groups have been most extensively examined in the electron-deficient heterocycles (pyridines, pyrimidines) because the use of a combination of ultraviolet and ionization constants in comparing the methylated and unmethylated tautomers was, and remains, a very powerful method of determining the tautomeric form[12].

Charge transfer effects have been extensively investigated in the case of proteins as well as for aromatic and heterocyclic compounds, and for those molecules, for example tetracyanoethylene, which produce strong charge transfer effects in the presence of donor molecules[15].

Where there are strong interacting groups present, such as nitro and amino groups together, leading to intramolecular charge transfer, new features arise. In synthetic and mechanistic organic chemistry, ortho and para positions tend to be similar, whilst the meta is different. But with these spectra it is the ortho and meta which tend to be similar whilst the para has a stronger long wavelength band shown in Fig. 6.8[16].

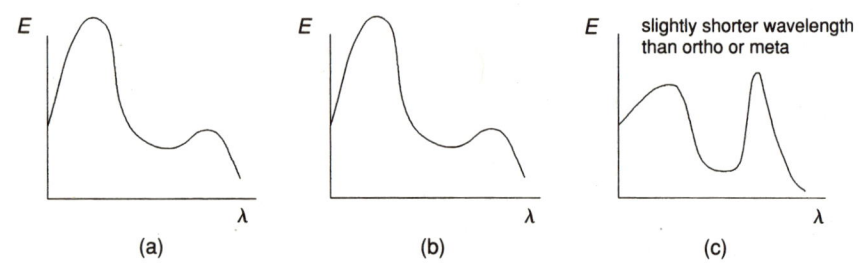

Fig. 6.8 *Comparison of spectra of (a) ortho, (b) meta and (c) para disubstituted benzenes with interacting substituents.*

These and other rules of substitution were originally formulated in the nineteenth century as the 'colour rules' in connection with dyestuff research[17]. Many synthetic dyes are combinations of a limited number of substituents ($NH_2$, OH, N=N, COOH, etc) on a limited number of aromatic frameworks (phenylmethane, anthraquinones, azo dyes from naphthalene sulphonic acids, etc) and these rules proved useful in predicting colours. In more recent times they have been put on a theoretical basis[16–19].

## 6.3 Summary

It is clear that electronic absorption band positions and intensities have a systematic basis in a wide variety of compounds. It is often possible to extrapolate literature values of absorption maxima of suitable reference compounds to a system under investigation. This can be done even where cross conjugation or polarization effects contribute strongly, although such cases require a more detailed theoretical basis of understanding of the origin of electronic spectra. There is further information in that the bands themselves are envelopes of bands from individual transitions, but this information is often only accessible at low temperatures. The contemporary problem for ultraviolet and visible spectroscopy is, however, that where compounds with novel conjugated systems are prepared as synthetic goals, the electronic spectra are often no longer recorded.

## References

1. Zechmeister, L. (1962) *Cis-trans Isomeric Carotenoids, Vitamin A and Arylpolyenes*, Springer, Vienna.
2. Miller, F.A. (1953), in *Organic Chemistry*, Vol III, (edited by H. Gilman) Wiley, New York, p. 168.
3. Fieser, L.F., and Fieser, M. (1959) *Steroids*, Reinhold, New York.
4. Grinter, R., and Threlfall, T.L. (1981) *UV Group Spectrometry Bulletin*, **9**, 106.
5. One student book where the rule for the spectral consequences of steric hindrance and of *cis* and *trans* double bonds in polyenes is correctly stated is Pavia, D.L., and Lampman, G.M., *Introduction to Spectroscopy*, Saunders College, Philadelphia, 1979.
6. Hamer, F. (1964) *Cyanine Dyes and Related Compounds*, Wiley Interscience, New York.
7. Fabian, J., and Zahradink, R. (1989) *Angew. Chem. Int. Ed. Eng.*, **28**, 677.
8. Mabry, T.J., Markham, K.R., and Thomas, M.B. (1970) *The Systematic Identification of Flavonoids*, Springer, Berlin.
9. Goto, T., and Kondo, T. (1991) *Angew. Chem. Int. Ed. Eng.*, **30**, 17; Eugster, C.H. and Märkli-Fischer, E. (1991) *Angew. Chem. Int. Ed. Eng.* **30**, 654.
10. Kuhn, H. (1958, 1959) The electron gas theory of the color of natural and artificial dyes. *Fortschr. Chem. organ. Naturstoffe*, **16**, 169–205; *ibid.*, **17**, 404–51.
    Zeichmeister, L. (1960) *Fortschr. Chem. organ. Naturstoffe*, **18**, 223–349. (Both these papers are in English.) Carotenoids are particularly discussed.
    Farrell, J.J. (1985) *J. Chem. Ed.*, **62**, 351–52.
11. Doub, L., and Vandenbelt, J.M. (1947) *J. Amer. Chem. Soc.*, **69**, 2714.
12. Albert, A. (1968) *Heterocyclic Chemistry*, 3rd edn. Athlone Press, London. pp. 376–419.
13. Stevenson, P.E. (1964) *J. Chem. Ed.*, **41**, 234–39.
14. Beavan, G.H., *et al.* (1955) *J. Chem. Soc.*, 2708.
15. Foster, R. (1969) *Organic Charge-Transfer Complexes*, Academic Press, London.
16. Grinter, R., and Heilbronner, E. (1962) *Helv. Chem. Acta*, **45**, 2496. This is discussed in J.N. Murrell (1963) *Theory of Electronic Spectra of Organic Molecules*, Methuen, London.
17. Berry, R.W.H. (1987) *Chem. Brit.*, **23**, 210.
    Griffiths, J. (1976) *Colour and Constitution of Organic Molecules*, Academic Press, New York.
    Griffiths, J. (1986, 1987) *Chem. Brit.*, **22**, 997–1000; **23**, 742.

## Further reading

The most useful of the available texts on organic electronic spectroscopy are probably:

Beavan, G.H., and Johnson, E.A. (1961) *Molecular Spectroscopy*, Heywood, London.
Jaffe, H.H., and Orchin, M. (1962) *Theory and Applications of Ultraviolet Spectra*, Wiley, New York.
  In many ways the most comprehensive book, but suffers from an inadequate index, deceptive chapter headings and an uneasy meshing of theoretical and descriptive material.
Morton, R.A., (1975) *Biochemical Spectroscopy*, Adam Hilger, London.
  Has an extensive list of literature references.
Murrell, J.N. (1963) *Theory of Electronic Spectra of Organic Molecules*, Methuen, London.
  This is a good book on the theory but may be found daunting to those wishing merely to pick up theoretical reinforcement to an understanding of spectral regularities.
Platt, J.R. (1964) *Free Electron Theory of Conjugated Molecules*, Wiley, New York.
Rao, C.N.R. (1974) *Ultraviolet and Visible Spectroscopy, Chemical Applications*, 3rd edn, Butterworths, London.
  Contains useful discussion and data, but as a reviewer has said: 'It is too short to do justice to the breadth of material it tries to cover'.
Sawicki, E. (1970) *Photometric Organic Analysis*, Part I, Wiley-Interscience, New York.
  A vast collection of information, with many tables of data.
Scott, A.I. (1964) *Interpretation of the Ultraviolet Spectra of Natural Products*, Pergamon, Oxford.
Stern, E.S. and Timmons, C.J. (1970) *Introduction to Electronic Absorption Spectroscopy in Organic Chemistry*, Arnold, London.
  Gives a coverage of much material missed by Jaffe and Orchin, for example the work of Braude's school, but to make up for it ignores a lot of other areas.

None of the books above have picked up the vast amount of systematic spectral explanation and information available in monographs and specialist texts such as:

Albert, A. (1984) *Heterocyclic Chemistry*, 3rd edn. Athlone Press, Chapman & Hall, London.
Goldschmid, O. (1971) in *Lignins* (edited by Sarkanan K.V.), Wiley Interscience, New York.
Goodwin, T.W. (1976) *Chemistry and Biochemistry of Plant Pigments*, Academic Press, New York.
Griffiths, J. (1971) *Colour and Constitution of Organic Molecules*, Academic Press, New York.
Sturmer, D.M. (1977) *Synthesis and Properties of Cyanine and Related Dyes, in The Chemistry of Heterocyclic Compounds*, Vol. 3, (eds. A. Weissberger and E.C. Taylor) Wiley, New York.

There are several compilations of spectra. The most extensive are:

*UV Atlas of Organic Compounds*, Vols 1–5, Butterworths London and Verlag Chemie, Weinheim 1966.
*Organic Electronic Spectral Data*,(1960) (Ed. M.J. Kalmet and others), Wiley-Interscience, New York, 1960 onwards.

Lang, L. (1966 – ) *Absorption Spectra in the Ultraviolet and Visible Region*, Akademiai Kiado Budapest.

Inorganic spectroscopy or at least transition metal spectroscopy which dominates this field is perhaps better served. Unfortunately, it is a subject which is difficult to tackle with a theoretical background, and many sound accounts in inorganic texts are so compressed as to be confusing or daunting to the neophyte. On the other hand there are several modern and comprehensive books:

Lever, A.B.P. (1984) *Inorganic Electronic Spectroscopy*, 2nd ed., Elsevier, Amsterdam.
    Lists the relevant literature.
Orgel, L.E. (1960) *An Introduction to Transition-Metal Chemistry, Ligand Field Theory*, Methuen, London.
    Provides a non-mathematical introduction.
Schläfer, H.C., and Gliemann, (1969) *Basic Principle of Ligand Field Theory*, Wiley, New York, Bibliography: pp. 500–19.
    Lists the relevant literature.

# 7 Colour

## 7.1 Colour perception

Colour, when considered in terms of one-to-one correspondence between a particular wavelength of light and a colour sensation (for example radiation of wavelength 450nm is commonly considered 'blue')[1], appears to be a straightforward phenomenon.

| violet | blue | green | yellow | red |
|---|---|---|---|---|
| 400 | 500 | 600 | | 700 |

In fact both the generation and the sensation of colour are complex. There is a wealth of mechanisms which give rise to light of a particular wavelength or range of wavelengths. The subtitle of Nassau's book[2], 'The 15 causes of color' may warn of the complexity of the subject. That white light can be split into a whole series of coloured lights, which we call a spectrum, and these can be recombined to re-form white light also reveals something of the complexity of colour perception.* The intricacy of the reception of colour by the eye is matched by the subtlety of the transmission of the resulting signals to the brain[3]. In addition to these physical aspects, there are strong psychological overtones which can be revealed by colour illusion experiments[4].

The vast majority of the colour we see is not single wavelength emitted light. The light from a sodium lamp is (almost), but that from an apricot is not. The eye cannot distinguish between yellow light, and white light from which the blue component is removed. For the same reason it cannot distinguish between the yellow of the sodium lamp, and the yellow light from a small prism which consists of a whole range of wavelengths. Whatever the complexity of the light, the eye can only distinguish in it one hue[5]. Indeed, light of the particular wavelength band corresponding to the hue observed does not even need to be present (Fig.7.1(d)).

This is nothing to do with colour optical illusion; this is the normal behaviour of the eye. However because the eye is so readily deceived by colour contrast, there is a need for objective instrumental colour measurement. To most people, colour is one of the most dramatic manifestations of the external world. It has

---

*Although Newton is commonly credited, and not only by the lay public, with the discovery of these phenomena, it is clear that Hooke and others had already observed the formation of the spectrum and its recombination by a prism. Newton's achievement was in the explanation of the observations. D. Thorburn Burns (1987), *J. Anal. Atom. Spectrosc.* **2**, 343.

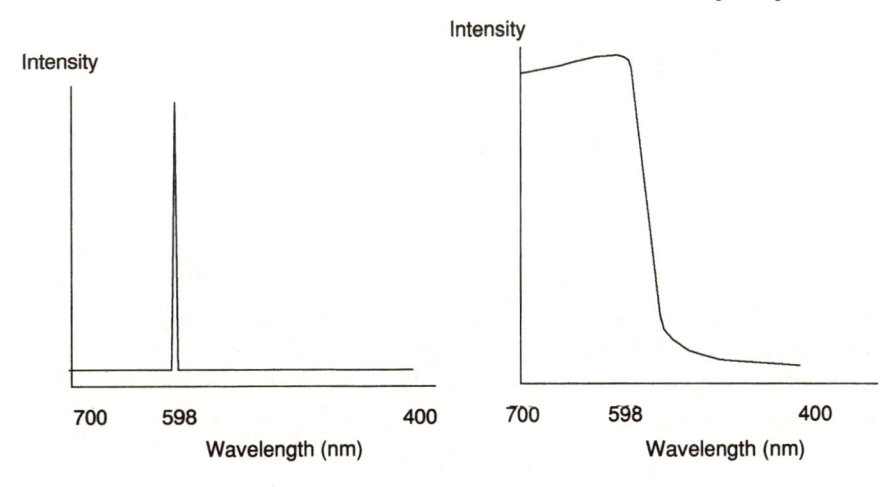

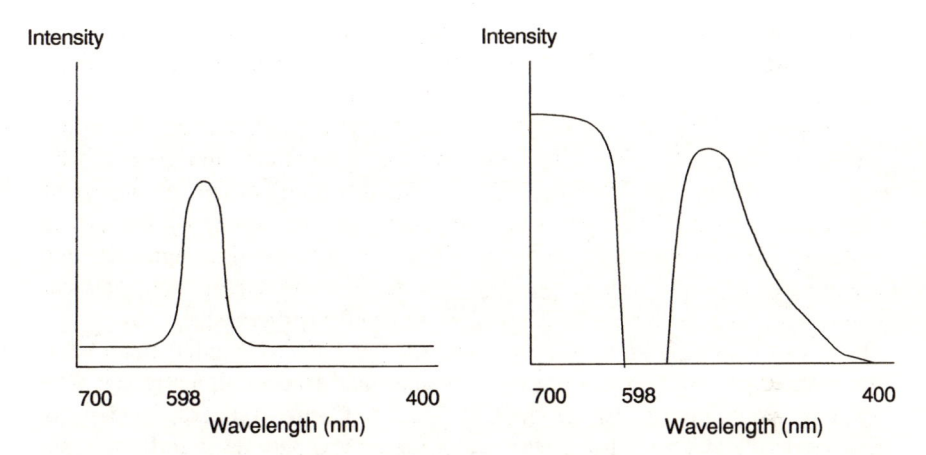

Fig. 7.1 *(a) Sodium lamp spectrum; (b) light reflected from typical yellow-orange object; (c) yellow-orange light passing through prism and slits centred on 598nm; (d) artificial spectral distribution yielding yellow-orange hue. This could be achieved from white lights by filters, or by mixing light from appropriate sources.*

come to play an important part not only in the traditional areas of textiles and paints[6], but in all areas of consumer products including both the packaging and the contents[7]. It is this inability of the eye to distinguish colours consisting of a spectral band from equivalent colours consisting of white light minus the complementary colour which leads to the confusion between additive and subtractive colours. Most of the colours we see are subtractive colours in which yellows plus blue will give green as in an artist's palette. By contrast, the addition of blue and yellow lights will give white light as in prism recombination experiments.

## 7.2 Colour vision theories

During the nineteenth century there were two theories of colour vision, one associated with Helmholtz in which it was asserted that the eye must have three receptors, because all colours, that is *hues*, could be built up by combinations of three lights; and one associated with Hering in which it was asserted there there must be receptors for four sensations because red, yellow, green and blue were perceived as distinct colours. Red and green form one contrasting pair (there is no such description as a reddish–green or greenish–red) whilst blue and yellow form the other.

It transpires that both are correct[8] and that the eye appears to have three receptors with the sensitivities shown in Fig.7.2. But the mechanism of the transmission of signals to the brain involves taking differences so that the brain actually processes one red/green signal and one blue/yellow signal, and the remaining parameter of the three is turned into the intensity factor.

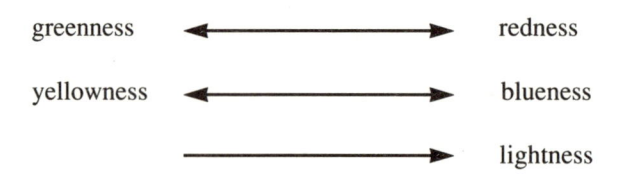

greenness ⟷ redness

yellowness ⟷ blueness

⟶ lightness

## 7.3 Colour measurement systems

These two representations can be used as the basis from which the two commonest colour measurement systems and their associated charts, the CIE tristimulus colour space[9] and the CIE $L^*a^*b^*$ (lightness, $a$-axis, $b$-axis) and CIE $L^*u^*v^*$ tabulations for surface colour and for lights respectively can be derived.

The relationship of these charts to the colour chip systems, illustrated for example in [6], will be perceived. Although still widely used, the colour chip comparison systems are being displaced by instrumental measurement, particularly now by spectrophotometric instrumentation, because the subjectivity of the comparison systems is avoided. There are other terms such as ANLAB, CMC which are increasingly encountered. These refer to new internationally agreed modifications based on improved colour metrics and measurements[10].

### 7.3.1 *CIE Tristimulus colour space*

Any colour on the original CIE system can be represented as two coordinates on the 'triangle', shown in Fig 7.3, which is distorted because of the second/(or negative) lobe of the receptor shown in Fig.7.2.

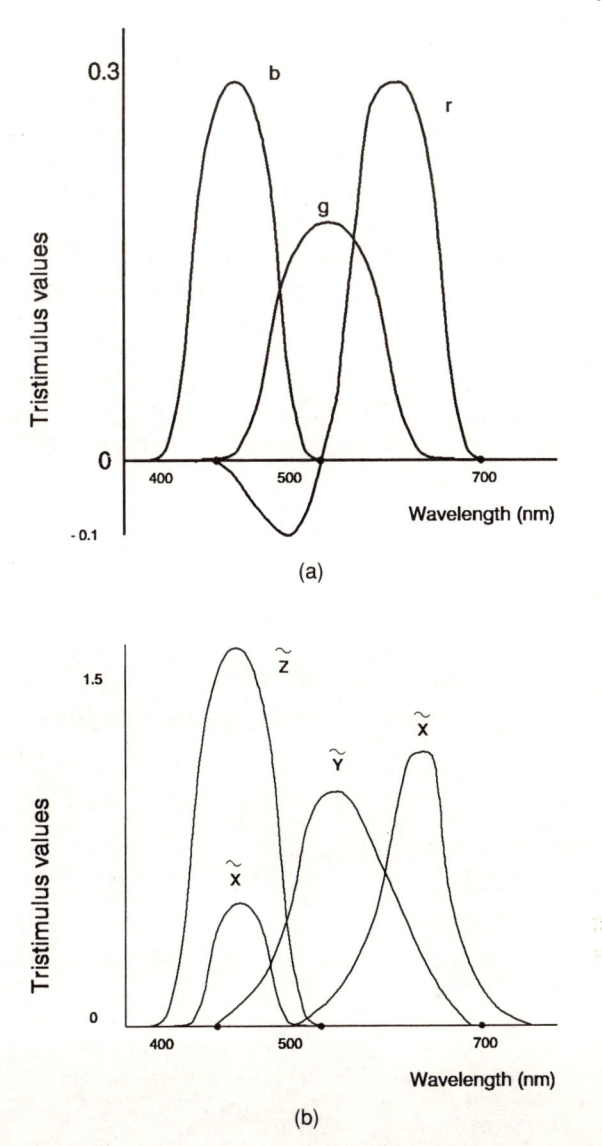

Fig. 7.2  *Tristimulus (a) response curve, and (b) after mathematical transformation.*

The whites are around the centre of the plot. Travelling out towards the periphery more *saturated*, i.e. purer colours unmixed with white, are encountered. Travelling around the curve the different spectral colours, that is the *hues*, are encountered. The straight line at the bottom of the graph represents the non-spectral colours, that is the purples. The complement of any colour, which is that colour seen when a given light is removed from white light, is found by drawing a straight line through the centre white. Where then is the

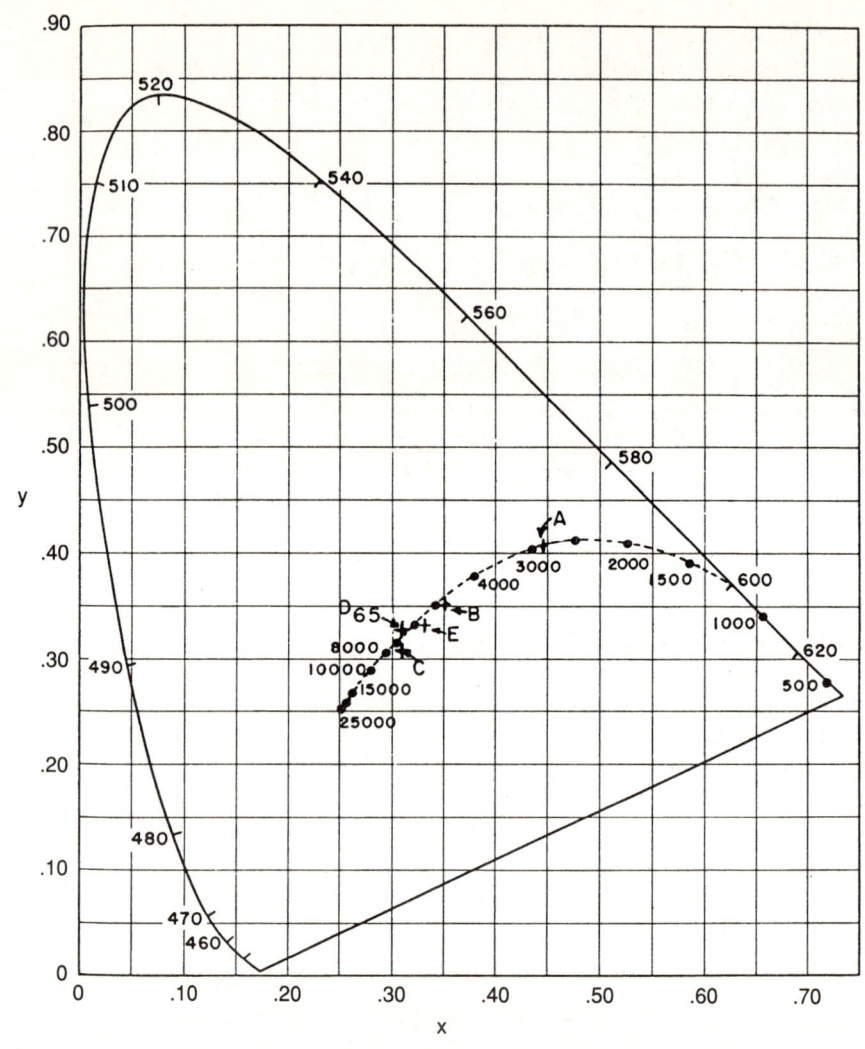

Fig. 7.3 *Maxwell colour triangle. CIE 1931 chromaticity diagram showing blackbody locus and CIE stimuli A, B, C, D65, and E. Reproduced from* Perception of Colour, *Evans (1974) with the permission of John Wiley & Sons.*

centre? That represents the illuminant, and so varies a little, depending on whether daylight, tungsten light, etc., is the illuminant. The curve across the centre of the diagram represents the colours of black-body radiation at different temperatures, and the adjacent letters represent standard, defined sources.

These directions, hue and saturation, are the same whether we are talking of lights or of complementary colours, i.e. those colours observed by reflection in which some of the wavelengths have been removed, for example by absorption. The third direction, the z-direction, is that out of the plane of the paper, and is the intensity direction. In the case of lights, intensity is straightforward: the sun is more intense at midday in summer than when veiled behind thin cloud in winter. In the case of surface colour this direction is the greyness scale, from black (absence of light) in the centre through to pure white.

## 7.3.2  L*a*b* representation

The $L^*a^*b^*$ or $L^*u^*v^*$ representation has conceptual simplicity, and furthermore the advantage that when the numbers are presented they can be immediately associated with the hue.

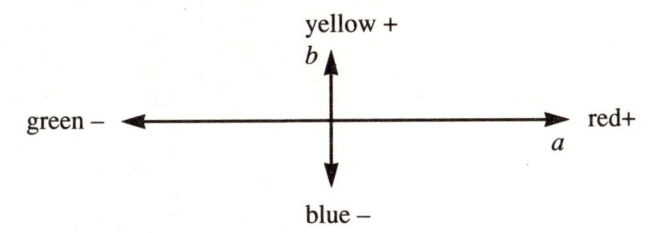

```
              yellow +
                 b
                 ↑
  green — ←——————+——————→ red+
                 |         a
                 ↓
              blue —
```

The third direction is again the intensity scale.

Although, as is clear from Fig.7.1, it is not possible to relate a given colour uniquely to a spectral distribution, the colour representations only have a meaning because a link can be made in the opposite direction. That is, a given spectral distribution can be assigned a unique number representing the colour. In order to convert the intensity of radiation between 380 nm and 760 nm, which are the limits of the sensation of colour for the human eye, into a number representing the 'colour', we need to return to the sensitivity chart of the eye. We simply multiply the measured intensity at each wavelength by the factor from the chart for that wavelength, and sum for each receptor in turn.

$$\text{Colour factor} = \sum_{380}^{760} a\tilde{X}, + \sum a\tilde{Y} + \sum a\tilde{Z}$$

$$= X + Y + Z$$

where $a_{380}....a_{760}$ are the measured intensities at each wavelength and $\tilde{X}, \tilde{Y}, \tilde{Z}$ are the weightings, that is, the sensitivity factors of the receptors taken from Fig.7.2, or better from tabulations[5]. In practice it is sufficiently accurate to record these values only every 10nm through the spectrum for all ordinary cases.

To plot out the $x, y$ coordinates of the CIE chart the two ratios are then calculated.

$$x = \frac{X}{X+Y+Z} \qquad y = \frac{Y}{X+Y+Z}$$

**Table 7.1**  *Abbreviated calculation based on the spectrum of Fig. 1(b)*

| Wavelength (nm) | Reflectivity(a) (arbitrary units) | $\tilde{X}$ | $\tilde{Y}$ (from tables) | $\tilde{Z}$ | $a\tilde{X}$ | $a\tilde{Y}$ (calculated) | $a\tilde{Z}$ |
|---|---|---|---|---|---|---|---|
| 410 | 0.3 | 0.38 | 0.01 | 1.80 | 0.114 | 0.003 | 0.540 |
| 430 | 0.3 | 2.33 | 0.10 | 11.37 | 0.699 | 0.030 | 3.411 |
| 450 | 0.4 | 3.72 | 0.42 | 19.62 | 1.488 | 0.168 | 7.848 |
| 470 | 0.5 | 2.12 | 0.99 | 14.00 | 1.060 | 0.495 | 7.000 |
| 490 | 0.6 | 0.33 | 2.14 | 4.79 | 0.198 | 1.284 | 2.876 |
| 510 | 0.8 | 0.95 | 5.13 | 1.61 | 0.760 | 4.104 | 1.288 |
| 530 | 1.0 | 1.69 | 8.78 | 0.43 | 1.690 | 8.780 | 0.430 |
| 550 | 2.0 | 4.27 | 9.80 | 0.09 | 8.540 | 19.600 | 0.180 |
| 570 | 3.4 | 6.95 | 8.68 | 0.02 | 23.630 | 29.512 | 0.068 |
| 590 | 4.0 | 8.61 | 6.35 | 0.01 | 34.440 | 25.400 | 0.040 |
| 610 | 4.1 | 8.50 | 4.27 | 0 | 34.850 | 17.507 | – |
| 630 | 4.0 | 5.06 | 2.09 | 0 | 20.240 | 8.360 | – |
| 650 | 3.9 | 2.15 | 0.81 | 0 | 8.385 | 3.159 | – |
| 670 | 3.8 | 0.68 | 0.25 | 0 | 2.584 | .950 | – |
| 690 | 3.8 | 0.15 | 0.05 | 0 | 0.570 | .190 | – |
| | | | | | $\Sigma = 139.25$ | 119.54 | 23.68 |
| | | | | | $= X$ | $= Y$ | $= Z$ |

It is tedious rather than difficult to do by hand, but trivial by computer. From the sums in Table 7.1 we have

$$X + Y + Z = 139.25 + 119.54 + 23.68 = 282.47$$

Thus, using the above expressions for $x$ and $y$:

$$x = 0.49 \text{ and } y = 0.42$$

This corresponds to a dominant hue of about 580 nm, as may be found from a chromaticity chart.

## 7.4  Problems in colour measurement

In practice the measurement of colour raises a number of problems. It is necessary to use an integrating sphere to collect all the light because of the nature of the surfaces. For example, a gloss surface tends to reflect specularly. It is necessary to measure the intensity of the reflected light, not the absorbed light that analysts are usually interested in. Over the years there have been several slightly different weighting schemes based on different angles of observations, different average observers, different illumination systems, etc., so it is important not to transfer information between different instruments without being sure.

The colour space about the chart is non-uniform in respect of spectral distinctions so that difference in coefficients between two samples may represent a distinguishable difference in one part of the chart but not in another[11].

In the end, it is an observer who must judge the matching in practice and for example, whatever the instrument records, a piece of velvet in not going to match a gloss paint exactly because the textures are so different[12]. This has caused particular problems recently with coloured representations on colour monitor screens, which are perceived significantly differently to reflection colours. The question of colour measurement in the presence of fluorescence became of commercial importance when fluorescent fabric whiteners were introduced[13]. Surfaces with colour-play effects, whether natural, as in opals[14], or products of modern technology, present a particular challenge in terms of colour description.

Colour measurement can also be applied to solutions. In this case it is necessary to calculate on the basis of the light transmitted rather than on the light absorbed, so the transmittance scale of a spectrometer, rather than the absorpt-ivity scale is necessary. The screening of indicators is a particular example[14]. The eye can distinguish a colour change involving a difference from white more easily than one involving an equivalent change of hue, say from yellow to orange. If, therefore, we can add a second dye which shifts the colour change across the grey point or white point then the indicator will be much more sensitive.

## 7.5   Summary

Increasingly, the ability to define and measure colour is leading to requirements to apply colour measurement ever more widely, for example in defining colours of pharmaceutical products. In addition, spectrometers with appropriate software are being used increasingly for colour measurement in place of tristimulus colorimeters. It behoves the analyst, therefore, to acquire a basic understanding of this mixture of physics, psychology and experience which is called colour and its measurement.

## References

1   Newton, I.(1730) 'The Rays are not Coloured', in *Opticks*, 4th edn, Reprinted 1952, Dover, New York. See also [4].
    Title of chapter in *Introducing Colour* (1975), Society of Dyers and Colourists, Bradford.
    Title of book of essays by Wright, W.D. and Hilger, A.(1960) London
2   Nassau, K., (1983) *The Physics and Chemistry of Color*, Wiley, New York. This work contains a useful list of general references.
3   Goldstein, E.B., (1989) *Sensation and Perception*, 3rd edn, Wadsworth, Belmont, California.

4 Brou, P. *et al.* (1986) *Scientific American*, 255 Sept. 80.
  Land, E.H., (1977) *Scientific American*, 237 Dec. 108.
  Wright, W.D. (1969) *The Measurement of Colour*, 4th edn, Adam Hilger, London Plate 6.
5 Chamberlain, G.J. and Chamberlain, D.G., (1980) *Colour; its measurement, computation and application*, Heyden, London.
6 McLaren, K., (1986) *The Colour Science of Dyes and Pigments*, Adam Hilger, Bristol.
7 Judd, D.B. and Wyszecki, G., (1975) *Color in Business, Science and Industry*, 3rd edn, Wiley, New York.
8 For some reason many physicists ignore the opponent colour theory, despite the fact that the research which substantiated the details of colour opponent cells led to Nobel prizes. See Goldstein, E.B.[3].
9 Crane, E.T., (1980) *Colour Measurement*, Pye Unicam, Cambridge.
10 Kornerup, A., and Wanscher, J.H. (1989) *Methuen Handbook of Colour*, 3rd edn, Methuen, London.
11 MacAdam, D.L., (1981) *Colour Measurement*, Springer, Berlin.
12 Hunter, R.S., (1975) *The measurement of appearance*, Wiley, New York.
13 Darragh, P.J., Gaskin, A.J. and Sanders, J.V., (1976) *Scientific American*, 234 April, 84.
14 Reilley, C.N., Flaschka, H.A., Laurent, S. and Laurent, B., (1960) *Anal. Chem.*, **32**, 1218.

# 8 Liquid chromatographic detection for multi-component analysis

## 8.1 Introduction

In chromatography there is a continual search for more sensitivity as well as selectivity from an assay method. In terms of detection many of the procedures rely on an ultraviolet–visible detector which although very efficient, in single channel mode generally lacks characteristic qualitative information and is rather insensitive to trace components. As a result there have been a number of developments which have partially addressed these shortcomings.

### 8.1.1  *Thin layer chromatography*

In thin layer chromatography (TLC) improved selectivity has evolved from high performance TLC plates, two-dimensional chromatography and radial TLC. These techniques have been particularly beneficial for compounds such as steroids in biomedical and pharmaceutical samples, where close chromatographic response is often found.

In terms of enhanced detection sensitivity and to aid quantitation, TLC plate scanning through densitometers operated in absorption, reflectance and fluorescence modes are now relatively common and can involve quite sophisticated technology. The detectors are primarily based on scanning linearly or through a preselected pattern across the TLC plate, with a light beam, of slit size which is adjustable for width and length. Mercury vapour, deuterium and tungsten lamps are commonly utilized, which give a spectral range from 190–800 nm. Detection limits, particularly with luminescence detection (in one case with laser technology) are generally in the low nanogram or picogram range. Multiwavelength data is also available from these detectors, either by a fast scanning process or through the use of a linear photodiode array multichannel detection mode. In both these formats large data matrices are possible and data handling and manipulation software is generally required. The major advantage of multiwavelength data is in spectral comparisons and the application of deconvolution techniques such as absorbance ratio to test the degree of resolution of components on the plate. Overall these densitometers can be expensive, but presently they are the most reliable method of obtaining quantitative information from a TLC plate.

## 98  UV detection

Other promising procedures for the evaluation of a TLC plate include developments in image processing with television cameras [1]. Speed of use is their great asset but reliable illumination of the TLC plate can be problematic and has still to be resolved for consistent operation.

### 8.1.2  Gas liquid chromatography

In gas liquid chromatography (GLC) developments have included improvements in selectivity through the use of capillary columns such as: support-coated-open-tubular (SCOT), porous-layer-open-tubular (PLOT), wide-bore fused silica (FSCOT) or glass or small-bore (0.25 mm i.d.) capillary columns. In addition to improve selectivity, mixed stationary phases [2] and coupled column [3] applications have been reported.

For detection involving spectroscopic systems, developments are to be found in a number of inventive detection modes. Typical examples are the flame photometric detector [4] which operates on the principle of chemiluminescence measurements from the decay of excited sulphur or phosphorus containing species by heating in a hydrogen–oxygen flame. These species give out chemiluminescent emission which can be detected by a conventional photomultiplier tube. Chemiluminescence detectors are also used in their own right. A typical example is the response from reduction/oxidation reactions obtained by passing over a catalyst the output from a capillary GC column and nitrogen dioxide and ozone, which leads to a chemiluminescent product when considering alcohols, aldehydes and ketones, sulphides and thiols [5].

A further spectroscopic detector is the photoionization detector which is based on the measurement of ionization current across a pair of electrodes, developed under the influence of a UV–lamp. Measurement of a number of groups of compounds, such as aldehydes and ketones and amines can be carried out and the method has the advantage of low background interferences. This can lead to lower detection sensitivities for compounds such as the barbiturates and benzodiazepine drugs [6], over the equivalent flame ionization detection procedure.

Additionally there are developments in hybrid spectroscopic detectors for GLC, through the high sensitivity of mass selective detection (in mass spectrometry) by using selective ion monitoring (SIM), Fourier transform infrared and atomic spectroscopy.

### 8.1.3  Separations of enantiomeric compounds

One very exciting field of analytical science over the last decade, is the separation of chiral compounds. Methods for chromatographic resolution of compounds with one or more chiral centres by TLC, GLC and high performance liquid chromatography (HPLC) are now well established, but detection of the individual isomers still provides problems. Conventional detection modes are acceptable where the separate enantiomers are available. However for detection and identification of the individual isomers the situation is not as clear-cut.

Detection regularly used in this context, includes polarimetric detection, which is based on the discriminating velocity caused by the optically active compound against the left and right components of circularly polarized light and circular dichroism detection, which is the differential absorption of the components of the circularly polarized light. There is however a major drawback in the overall sensitivity of detection. Both methods rely on having large optical rotations from relatively high concentrations and cell volumes and in many cases and particularly with drugs the former position is not regularly found. By the very nature of GLC and HPLC, concentrations and sample sizes (flow cell volumes in HPLC) are necessarily small. These difficulties are magnified when using chiral stationary phases in HPLC, as the density of surface coating of the immobilized chiral selector on the phase is generally low and small sample volumes and concentrations have to be introduced onto the column packings to prevent overloading. With these detection modes, laser-based systems allow one to work at improved limits of detection but often this is still not low enough to detect trace levels of an enantiomer.

In practice in HPLC and TLC, for absolute confirmation of the enantiomeric identity, circular dichroism (CD) is the most suitable methodology. This is an absorption technique which distinguishes between right and left circularly polarized light, with the possibility of examination of the area around the chiral centre. But unfortunately the values obtained from the CD spectrum are considerably lower than the respective absorption effect, and for efficient operation, off-line determinations, after mobile phase elimination, yields the best results.

## 8.2 High performance liquid chromatography

The major analytical developments over the last 15 years have been in the technique of high performance liquid chromatography (HPLC). Interests in progression in HPLC are still very prominent even though there is considerable growth of developments in techniques such as capillary electrophoresis, supercritical fluid chromatography, and a certain revival of interest in gas chromatography.

In HPLC, selectivity and sensitivity for trace level, multicomponent mixture analysis of chemical, pharmaceutical and biochemical mixtures has been addressed in a number of ways. These span the areas of miniaturization, of the column packing material, through to the pre- and post-column reaction chemistries and to the detection mode and sensitivity.

Detection in HPLC is based on the fundamental approaches of measurement of the solutes through detection of a bulk property of the solute and mobile phase, a specific property of the solute and on detection after removal of the mobile phase.

The most prevalent detectors are spectrophotometrically based (falling into the second category), which have the advantage that they are generally stable,

## 100 UV detection

have wide applicability and are inexpensive. Initially a simple fixed wavelength UV-detector format was used at 254 nm (using a phosphor convertor, 280 nm is possible), which routinely utilize a low-pressure mercury source, giving high energy output at discrete wavelengths. The energy from the lamp is passed through a narrow bandpass filter and directed on to a flow cell with a collimating lens before detection by a photocell (Fig. 8.1a). Flow cell design is usually of a 'z' form for the inlet and the outlet (at different sides of the cell). This design allows about a 10 mm pathlength for the cell. In the early cells, the cell volume was between 15–25 $\mu$l, which gave relatively broad chromatographic peaks. Later cell volumes were reduced to 8 $\mu$l, with subsequent improvements in peak shape, and these cells are still commonly used today.

The fixed wavelength detector has been used in many applications, but there are a number of compounds which have insignificant absorbance at 254 nm, typical examples being the barbiturate drugs and many pesticides. Following from the fixed wavelength, the next development utilized variable wavelength detection and therefore addressed the above problem by allowing measurement detection at the wavelength maximum. The mercury lamp was replaced with a deuterium or tungsten lamp, to cover the range 190–650 nm. For monochromation and detection, a manually adjustable diffraction grating, which has a wide bandpass and a photomultiplier detector is typically used (Fig 8.1b). This geometry gives a stable response, low noise level and consequently high signal-

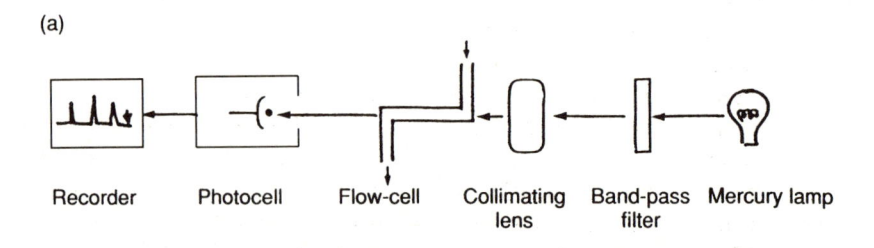

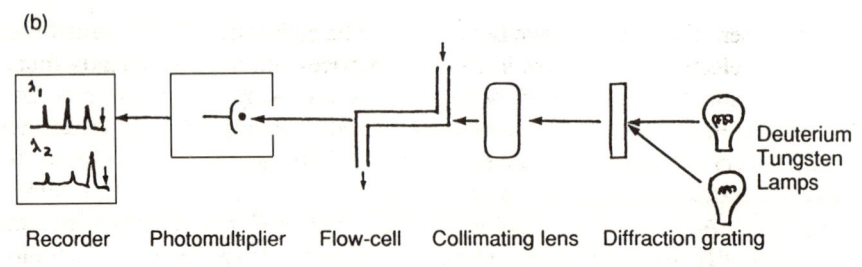

Fig. 8.1 *Layout of (a) fixed wavelength detector, and (b) variable wavelength detector.*

to-noise values and linear response over a wide absorbance range, according to Beer's Law. This stability typically allows the variable wavelength detectors to be used down at 0.001 absorbance units, although to achieve a relatively low baseline noise in commercial instruments the cell volume is raised to 15 $\mu$l. In the most recent detectors, a reference cell has been incorporated, which can use a beam splitter principle from the lamp or a method not strictly double-beam based, where the reference cell is measured by a photocell.

Although these spectrophotometric detectors are 'workhorses' in the analytical laboratory, the format of the detector does not yield information about the identity of the analyte of interest. This can only be gained if a pure sample of the expected compound is chromatographed under identical conditions. Additionally as regards the range of detection and sensitivity, many compounds do not possess a suitable chromophore and therefore detection by this mode is not feasible.

### 8.2.1  Universal detectors in HPLC

Much of the need for progression towards improvements in detection selectivity and sensitivity, result from the absence of suitable stable and sensitive universal detectors in HPLC and this has led to a number of alternative possibilities being explored. There are however a number of universal detectors available, which are bulk property based and utilize the conductivity, refractive index (RI), dielectric constant, density, electron capture, flame ionization and heat of absorption properties of compounds. But for general operation these detectors can suffer from a number of drawbacks, of which the detection stability is the main concern. Developments in this field are however continuing and there have been sensitivity improvements in refractive index detection and other detectors, as reviewed by Yeung [7]. In RI mode these sensitivity improvements have been achieved through reduction of fluctuations in the mobile phase RI, by temperature, flow rate and mobile phase composition stabilization. Additional innovative developments have allowed extensions to refractive index gradient detection [8] and the introductions of techniques such as spectral interference refractometry involving diode array spectrophotometry [9].

### 8.2.2  Electrochemical detection

Electrochemical detectors have been shown to be both selective and sensitive for levels of electroactive species in complex matrices and have particularly found favour with the clinical and neuro-chemist for assay of amino acids and catecholamines in biological samples [10]. Unfortunately the detector has not been universally accepted into the analytical laboratory, possibly due to the adverse initial design problems, which affected the operation of the working electrode. Amperometric, coulometric and voltammetric detectors are all commercially available from a number of sources. Amperometric and coulometric are the most common and although they differ considerably in their mode of operation, they only differ slightly in their detection sensitivity. (In the amperometric detector only a small percentage of the analyte is utilized in detection, whereas 100% is converted in the coulometric detector.)

One interesting development in electrochemical detectors is the multiple electrodes in series in the 'Coulochem Electrode Array System' (ESA, Bedford, Massachusetts, USA). Each electrode can be set to the optimum potential for oxidation or reduction which is observed in three-dimensions. This extra dimension has been used in a ratio mode which is characteristic of the component under test.

Voltammetric detection is relatively new in commercial terms. It is based on the applied voltage to the working electrode being varied with time and the current response is measured as a function of both time and potential. This dual response results in a three-dimensional 'voltammogram' for each chromatographic peak (Fig. 8.2). This extra dimension can be utilized for identification purposes however, by scanning the applied voltage, although there can be the major drawback of the poor detection limits, (a decrease of around three orders of magnitude, against a regular electrochemical detector).

As suggested above, the electrochemical detector has not fully been accepted by the analyst, although it does offers considerable advantages in sensitivity. Applications are, however, constantly expanding, in particular for drugs, their related compounds and metabolites [11] and in environmental applications.

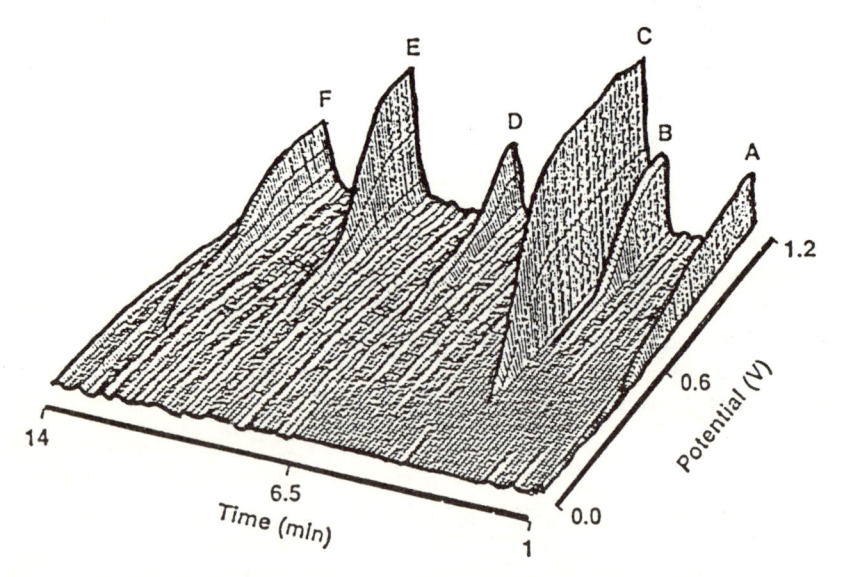

Fig. 8.2 *Three-dimensional voltammetric detection of current, time and potential. The separation of phenolic acids on a 5µm ODS column with a mobile phase of methonol: 0.1m ammonium phosphate pH4.0 (15:85 v/v). Flow rate: 1.0 ml min$^{-1}$; scan rate: 2.0 V s$^{-1}$; potential: –0.2V (Ag/AgCl). Chromatograph peaks: (A) gentisic acid, (B) vanillic acid, (C) caffeic acid, (D) 4-coumaric acid, (E) ferulic acid, and (F) sinapic acid.*

### 8.2.3 *Light scattering detectors*

Among the other interesting developments in the field of detectors for HPLC are the light scattering systems. The operation can occur in an evaporative mode using a nebuliser and conventional energy source, such as the tungsten lamp ['Mass Analyzer' ACS, Cheshire, UK]. Alternatively a laser source is used and a number of laser light scattering detectors are available. With both types of detector the nature of the solvent and the solute are important [12]. It is interesting to note the increase of use of lasers generally in commercial detectors over the last five years, which can be attributed to the improved stability of the technology. However small lasers operating in the short UV and visible wavelength ranges are not commonly available and are expensive, and the long wavelength operation (>800 nm) gives insufficient sensitivity for general applications.

### 8.2.4 *Conductivity detector for ion chromatography*

One specific 'universal' non-spectroscopic detector which has grown in popularity alongside the technique of ion chromatography is conductivity. The detector based on ionic species, requires an ionizing medium as the mobile phase. It is very efficient for monitoring charged species and low limits of detection are possible if care is taken in reducing baseline noise by replacing highly conducting eluents or reducing their concentrations [13]. Improvement in detection sensitivity is also achieveable if sample compression can be obtained on first introduction of the sample in solution. The basis of this is choice of sample solvent which can lead to depositing the sample at the top of the HPLC column.

### 8.2.5 *Multifunctional detectors for HPLC*

In addition to these single mode detectors improvements in selectivity through multifunctional detection has been explored. In the commercial world, Philips, Dupont and Perkin Elmer have all entered this market. In the case of the latter system UV, fluorescence and conductivity were all available as detection modes. However although the principle appears to be sound, in practice the single elements of these detectors are not very sensitive and therefore have limited application.

## 8.3 Hybrid detectors for HPLC

In the absence of the true 'universal detector' there have been considerable moves towards combining spectrometry with HPLC in the so-called 'hybrid systems'. Interesting, novel interfaces have been developed to provide the link between the chromatograph and techniques such as atomic absorption (AA), nuclear magnetic resonance (NMR), fourier transform infrared (FTIR), and mass spectrometry (MS). Problems of technique compatibility are obviously paramount in these systems, especially in pharmaceutical and biomedical analysis where reversed phase HPLC is most commonly evidenced and therefore the presence of aqueous mobile phases have to be considered.

### 8.3.1  HPLC–Fourier transform infrared (FTIR)

HPLC–FTIR coupling in normal phase operation in HPLC were initially considered in this area, as reversed-phase presented particular difficulties where water and alcohols absorb very rigorously in the infrared spectral region. However, elimination of these solvents has been examined in many instances, and in the simplest cases solvent evaporation or extraction of the solutes into an IR-compatible solvent is possible. But these procedures are somewhat crude for stable IR assay of the solutes and as a result a number of more innovative procedures have been developed [14]. One of these methods was suggested by Hellgeth et al. [15], where the eluent from the reversed-phase column is introduced with an organic solvent (such as deuterated chloroform) in a T-junction and leads to a segmented flow. The analytes are then extracted into the organic solvent in a mixing coil and the two phases can be separated by a membrane separator, prior to drying of the organic phase and passing into a suitable flow cell for running of the IR spectrum.

### 8.3.2  HPLC–nuclear magnetic resonance (NMR)

The coupling of HPLC to NMR can provide a very powerful analytical system, but the interfacing problems are even more acute and the choice between online and offline operation often falls with the latter format, especially when considering the problems of high acquisition rate required in HPLC. Flow cell development is considered the crux of the problem, where the flow rate, cell volume and sample tube spinning are all taken into account. But online procedures are carried out and either stop-flow scanning or continuous-flow detection using a flow NMR probe has been reported by a number of research and industrial groups [16]. Protonated solvent effects, and particularly the presence of water from reversed-phase eluents, can be dealt with through homogated decoupling methods [17].

### 8.3.3  HPLC–atomic absorption detection (AA)

The use of HPLC–AA is another interesting development and in its simplest form direct coupling to the nebulizer with capillary tubing has been successfully reported for the detection of trace metals in a variety of samples. A typical example was discussed by Hill et al. [18] where the presence of tributyltin was assayed in harbour water, where it is used in antifouling paint for small sailing craft. The direct result of using this paint was to cause sterility in shell fish which breed in these harbour waters. With a slightly more sophisticated interface (Fig. 8.3), Ebdon and co-workers [19] were able to assay tetraalkyllead

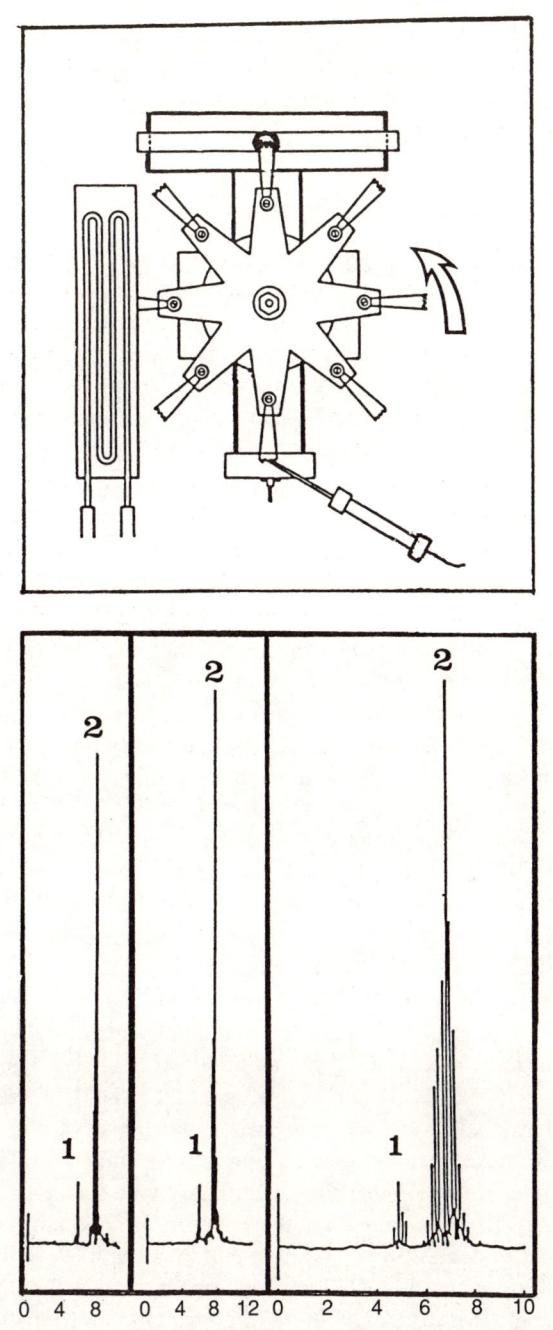

Fig. 8.3 *LC coupled to atomic absorption for the assay of tetraalkyllead compounds in petrol.*
*The stepper motor rotated interface carries the analytes and the mobile phase from the end of the LC column to the quartz tube/burner system. A heater then cleans the platinum sampling collector before its return to the end of the column. The column was 100mm × 4.6mm $C_{18}$ and the chromatographic peaks are tetramethyl lead; 2-tetraethyl lead.*

compounds in petrol with a rotating spiral interface to carry the eluant from the HPLC column to the AA burner head. As an extension to this mode of detection, coupling of HPLC to either a graphite furnace or inductive coupled plasma has also been reported [20].

### 8.3.4 HPLC–mass spectrometry (MS)

In the main, the interfaces discussed above have lacked commercial interest. This however has not been the position with HPLC–MS which is one of the most powerful hybrid systems. A number of successful interfaces are available, albeit at considerable cost. These have resulted from: incompatibility problems of large polar solvent flows (reversed-phase), involatile samples, presence of buffer salts and flow differences. These systems include the moving belt, thermospray, direct liquid injection, continuous fast atom bombardment and the most recently available commercial unit of electrospray. They all have their strengths and weaknesses in accordance with the samples to be analysed and the MS equipment to be interfaced. One of the first systems was the moving belt which was based on: depositing the column eluant onto a polyimide belt, solvent removal and volatilization of the sample into the ion source [21].

Presently the most favoured interface is the thermospray which can produce ions without a source of ionizing electrons and is generally operated in the CI mode. Its advantages include its ease of operation, (benchtop units are available from the Vestec Co., Houston, Texas, USA), the capability of using normal HPLC flow rates and the possibilities of obtaining spectra from different ionization modes which help to identify structures.

One of the most significant recent developments in this interface field is the electrospray (ES). The system is similar to the thermospray in operation; however, detection limits around those achieved for GLC–MS are possible, whereas with the thermospray, lower limits of detection are regularly found. Capillary columns are required (conventional columns and flow rates in thermospray) and flow rates of $1-100\,\mu l\,min^{-1}$. Although sensitivity is one of the advantages of the interface, the largest potential appears to lie within capability of handling large to macromolecules as multicharged ions. Quadrupole mass filter MS has most often been used in this field [22] although there are reports of coupling with sector MS and FT–MS [23]. Typical examples of the work with macromolecules have been reported by Van Berkel and co-workers [24] where molecular weights up to 66 000 (bovine albumin) have been examined.

There is therefore great potential in these interfaces although at present certain difficulties exist in dealing with complex unknown samples. This is due to the use of only mild ionization technique such as CI, except for the thermospray which uses EI, and therefore only limited fragmentation information is available for identification purposes. One solution to this is the coupling of MS–MS. Initially a soft ionization technique like CI or fast atom bombardment (FAB) is used and then collision induced decomposition can lead to structure determination. With these instruments which are available from a number of manufacturers (including VG Instruments and Finnigan) and the introduction of new software design strategies for experimental development and structure elucidation through expert systems, the use of HPLC/MS and HPLC/MS/MS is now a far more easily handled analytical tool.

In addition to these interface developments there have been a number of developments to improve the basic mass analyzer of the MS, through the magnetic sector, quadrupole, ion trap and time of flight modes [25].

## 8.4 Rapid scanning spectroscopic detection

Although the above tandem systems of HPLC–AA and HPLC–MS etc. have a large potential, the largest commercial success has resided with the rapid-scanning spectroscopic detector and especially the multichannel detector. These detectors have come in a number of forms: from stopflow scanning, oscillating grating based and vidicon systems, to charge-coupled devices, and the linear photodiode array detector, with varying degrees of commercial success.

### 8.4.1 *Luminescence multichannel systems*

Absorptiometric detectors have dominated this field, but it is interesting to note that there have been commercial developments in the area of luminescence multichannel detection. The first successes with commercial luminescence systems for HPLC were reported on using a commercial intensified linear photodiode array (LDA) detector by Clark [26] with application to a $\beta$-blocker drug and polynuclear hydrocarbons as illustrated by Fig. 8.4, (detector, Tracor Northern, Madison, Winsconsin, USA) and also application to these aromatics in the unintensified form [27] (Perkin Elmer, Beaconsfield, Buckinghamshire, UK).

With regard to the long awaited charge-coupled devices (CCD) in HPLC detection, there still appear to be some problems of short wavelength registration and stable wavelength resolution, although there are CCD spectrometers available from companies such as Biolink Technology (Biolink Technology Ltd., Cambridge, UK) and Photometrics (Photometrics Ltd, Tucson, Arizona, USA).

### 8.4.2 *Absorptiometric multichannel detection*

Apart from these more limited hybrid applications, the major mode of this form of detection involves the multichannel absorptiometric HPLC detector. This detector is commercially found in two formats, the linear photodiode array (LDA) on which the systems are primarily based, with over 20 derivatives from the majority of the primary manufacturers of HPLC instrumentation available, and a system based on an oscillating grating. This latter mode allows rapid determination of wavelength information by mechanical monochromator movements. Waters (Waters/Millipore, USA) and Barspec Ltd (Rehovot, Israel) have promoted this format and the system has been available for a number of years.

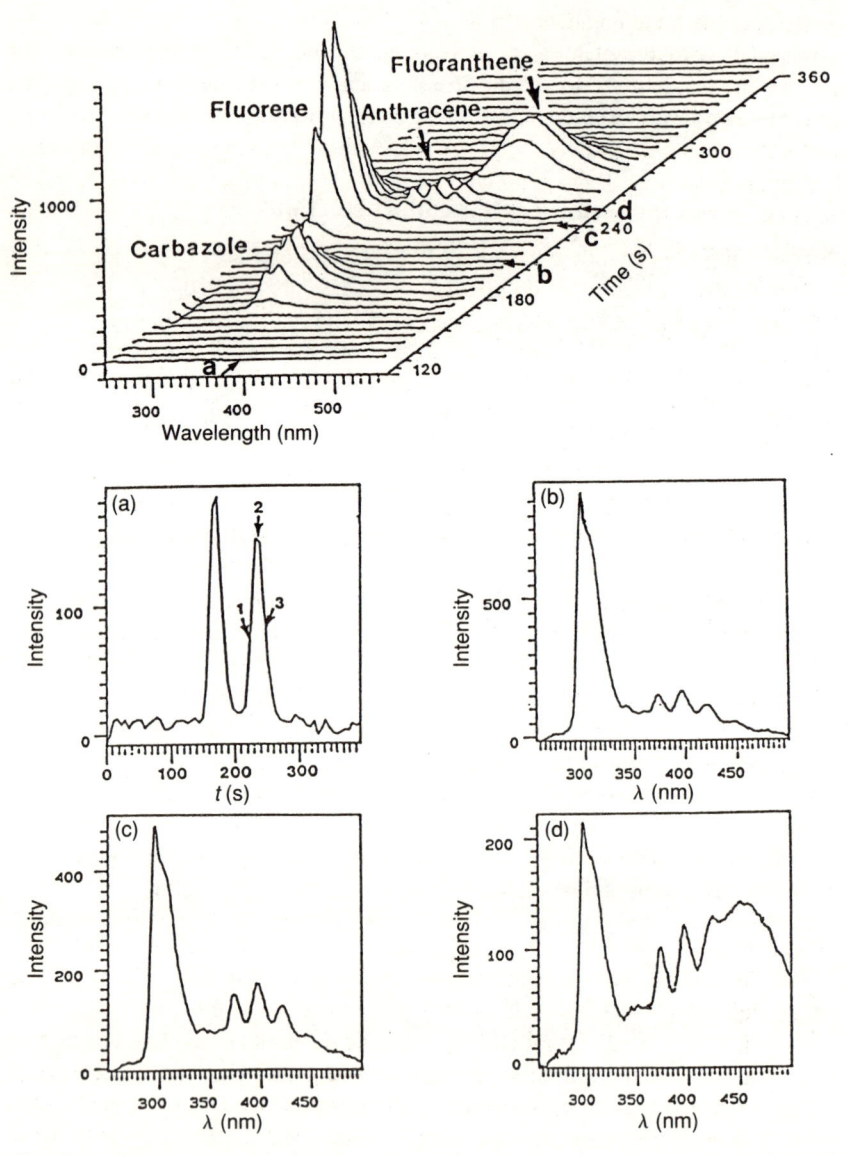

Fig. 8.4 *Presentation of three-dimensional, wavelength and time data from the luminescence linear photodiode array detector.*
*The usefulness of the extra dimension of fluorescence wavelength information is demonstrated by fuller examination of the chromatogram taken at 379 nm (a). Although the first peak is pure, the second contains the coeluting anthracene and fluoranthene, as illustrated by (b) at t = 225 s, (c) at t = 240 s, and (d) at t = 250 s.*

But as stated, the successes have been with the LDA mode and as commercial developments have formed, two divisions of the basic design, as originally introduced by Hewlett-Packard in 1981 (HP 1040A), are now marketed. One form consists of a detector which is fundamentally intended to capture data across the UV or visible wavelength ranges during the elution, and uses between 256–1024 diodes. A common terminology for the basis of this form of detection is 'self-scanned'. In terms of detection sensitivity, there is an approach towards the conventional variable wavelength single channel detector. A large matrix data can be generated and powerful computing capabilities are required to usefully interrogate this data. These large data matrices open up numerous possibilities for solute identification and checks on peak purity and homogeneity through data manipulation. But this can be very time-consuming and in many cases, particularly in the quality control environment, there is slight overkill in producing such a large data set. As a result, it is often suggested that this form of the detector is best employed in a research and development laboratory. Typical examples of this form are the Hewlett Packard, HP 1090M and the Waters 990.

In contrast the other format involves the 'stand-alone' detector which is generally more limited in its spectral capture capabilities but more slanted towards the regular monitoring position, and a number of commercial derivatives have been developed. These multichannel detectors are regularly based on a limited range of elements in the detector array (commonly around 35 diodes). With this reduction in the number of diodes in the overall detector format, where the optical geometry of the LDA size is fixed, it has been possible to increase the diode surface area. In turn this allows for longer integration time on the diode (used in conjunction with slower scanning speeds) and results in reduced signal-to-noise values (SNR). Noise is also reduced in this format by externally scanning the detector array. In the original detector, where a self-scanning format is used there is an amplification of the noise generated in the integrated circuit multiplexers. Therefore an improved detection sensitivity is given over the 'complete system', but this has to be traded off against reduced spectral capture capabilities or a wider bandwidth of detection. These detectors are commonly characterized by including an in-built microprocessor for acceptable limited manipulation of the data captured, although it is possible to extend their capabilities through down-loading the data onto a microcomputer.

Overall, however, the major advantage of the detector is the rapid acquisition of spectral data, many times a second, during the actual elution. The detector is formed from an array of light sensitive photodiodes, which are located in the focal plane of a diffraction grating polychromator. Each photodiode can be set to correspond to a narrow wavelength range and the resolution is determined by the overall geometry of the polychromator–LDA combination. In the first system introduced by Hewlett Packard, in 1981, the array consisted of 256 elements, which covered the range 190–600 nm. This gave a spectral resolution of ±2 nm (around 50 diodes were used for reference purposes). It is accepted that the resolution limit imposed by the LDA format can be a drawback, when compared with conventional spectrophotometers (0.25 nm resolution), but this has to be considered in the context of the components of interest, their spectral characteristics and the system format. As an example many components in biomedical and pharmaceutical samples give broad bandwidth spectra and therefore 2 nm resolution is not problematic. But when the more limited spectral array is used (i.e. 35 diodes) and complete or large parts of spectra are involved, then the resolution of around $\pm(5-10)$ nm can be problematic.

Apart from the number of elements in the detector array the primary difference between the formats lies in the method of scanning of the elements. In the initial system the array was self-scanned which does lead to some noise amplification from the integrated circuit multiplexers, whereas in the stand-alone units noise has been reduced through externally scanning the array.

From both these instrumental formats a large amount of data is generated, as illustrated by Fig. 8.5, and computer-aided algorithms have been developed which allow data reduction, manipulation and presentation of the data [28]. It is then possible with this extra dimension of spectral information to use these computer-aided methods to improve the information content from the chromatographic system. These span the development phase of an HPLC method; from method initiation and optimization, through to chromatographic peak purity determination and method validation through the use of a bench-top microcomputer and suitable software.

(a) *HPLC method optimization*

In the early stages of method LC development for a sample mixture, the achievement of acceptable resolution ($\alpha$) and analysis time are essential elements in the chromatographic response. Developing a separation is often not easy, particularly when the analyst is handling complex multicomponent samples. The often used trial and error approach to developing a method can waste time and is not very efficient and generally more structured optimization procedures to method design are commonly carried out.

These structured approaches exist as simultaneous or sequential methods, which essentially differ in the mode of reaching the final objective of the global optimum set of conditions. Thus they differ in the arrival at the best set of conditions and in the former method a predefined scheme is used to reach that optimum, whereas in the latter case search algorithms at each step direct the next experiment based on the separation achieved by the last experiment. These approaches are particularly suited to the LDA detector where a check on the progress of the method development through peak tracking and chromatographic peak purity and homogeneity can be helped by using the spectrochromatographic information from the multichannel detector.

Peak tracking is a particular problem in any optimization method where the chromatographic peak order can change. It is possible through spectral recognition algorithms to monitor peak positions. This can be carried out by

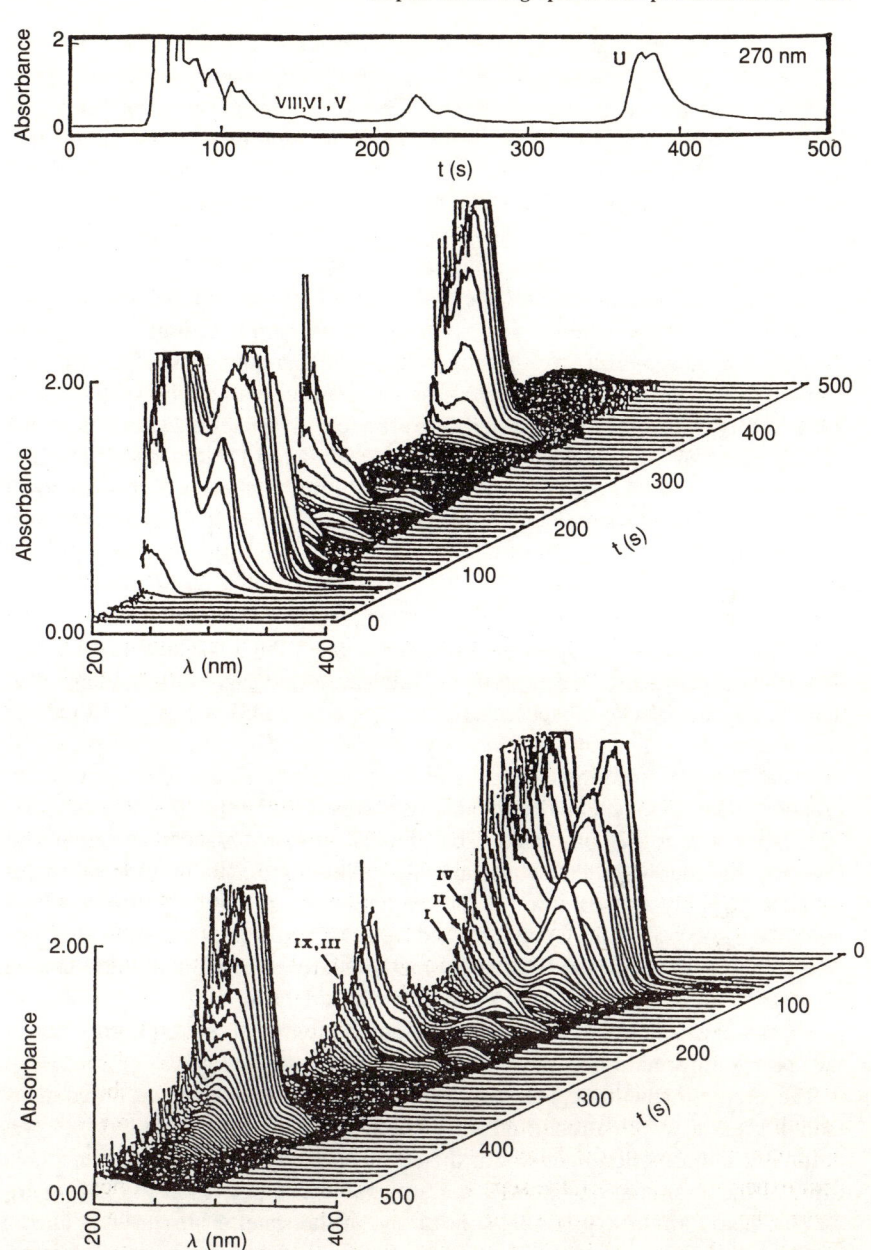

Fig. 8.5 *Isometric plot of the direct injection of a urine sample containing the antidepressant drug, zimeldine and its metabolites.*
*Illustrated is the single wavelength chromatogram at 270nm (top); the forward projection where the drug (I) and its major metabolites (II, III, IV and IX) are hidden by the endogenous background; the reverse projection showing the presence of the drug and metabolites.*

peak area measurements or by the extent of elution through total integral values [29], where the total sample is passed into a detector with the HPLC column removed and the response integrated. Then with the column in place the integrals of the individual peaks are totalled and compared. Throughout the method development the purity of a chromatographic peak is also essential and spectral information can be examined to aid peak purity assessments.

(b) *Spectral libraries*

It must be said however that UV spectra in HPLC are not ideal for identification purposes, especially when the spectra are broad bandwidth and the sample spectrum is susceptible to solvent effects. Particularly bathochromic and hypsochromic shifts can cause problems in data archiving along with temperature and type of organic modifier. Nevertheless with careful attention to the parameters in use, spectral libraries can be operated in LDA detection. The first use of libraries in this detection format was reported by Fell and co-workers in 1984 [30] and since then a number of commercial software packages have been released.

(c) *Chromatographic peak purity and homogeneity validation*

The extra dimension of spectral information from the LDA detector has not only, through computer aided methods, introduced the capability to identify an analyte by careful use of spectral libraries, but in addition has permitted the consideration of resolution efficiency of the HPLC method. This can be achieved through examination of the homogeneity and purity of the chromatographic peak (homogeneous peaks are not necessarily pure).

This latter aspect is of considerable interest in many branches of analytical research. In order to validate chromatographic peak purity and homogeneity it is possible to apply one or more of the many computer aided algorithms which exploit the spectral information acquired by LDA detectors. Numerous methods based on spectroscopic and statistical methods are available and these can be separated into two approaches of simple procedures and matrix (chemometric) mathematical methods to interrogate the chromatographic peak through use of the spectral information.

The simple mathematical approaches, which are primarily based on comparison of spectral information or on wavelength ratio data, can give a fast interpretation of the purity/homogeneity of the chromatographic peak. One of the most rapid is the examination of the spectra at numerous points on the chromatographic peak profile. By normalising to one of the spectra present, this allows comparison by overlaying. Of the other basic methods a number revolve around adaptations of absorbance ratioing, which has a long history in UV-visible and MS spectroscopy [31]. The method is best carried out by plotting the ratio of absorbances at two discrete wavelengths over the elution profile. For a pure component, the molar absorptivity ($\Sigma$) at wavelength $\lambda_1$ is directly proportional to that at any other wavelength $\lambda_2$.

$$\Sigma_{\lambda_1} = K_{1,2} \cdot \Sigma_{\lambda_2} \tag{1}$$

The constant $K_{1,2}$, for the pure component, is characteristic at the wavelengths used. If agreement with the Beer–Lambert law is given then this constant is independant of concentration and the absorbance values $A_1$ at $\lambda_1$ and $A_2$ at $\lambda_2$ are related through the same proportionality constant

$$A_1/A_2 = K_{1,2} \text{ at } \lambda_1, \lambda_2 \tag{2}$$

As a result the absorbance ratio between the wavelengths chosen should be constant and the graphical form shows as a flat-topped (square) waveform. When there is the presence of a partially co-eluting impurity peak, with a different $K_{1,2}$, then this difference is observed in the absorbance ratio and in the graphical presentation.

The major advantage of the AR method is its capability to give a rapid test for the presence of a co-eluting impurity. There are limitations and the method can suffer from some problems which relate to dissimilarity of the spectra, their relative absorbtivities and the overall lack of sensitivity of the method (approx. sensitivity to >10% w/w impurity). The AR method is generally used as a qualitative technique and numerous adjustments have been made to enhance its operation and to allow quantitation. These include a number of commercial initiatives, which usefully produce a single figure value to indicate purity. This is considered important, as single figures can be introduced directly into reports, can lend themselves to further numerical evaluation and offer the possibility of being incorporated into automated systems. Such systems include 'Absorbance Index' (Perkin Elmer) and 'Peak Purity Parameter' (Varian) which give considerable improvement upon the AR method. The absorbance index method is calculated over the upslope and downslope of the peak and regression analysis checks the linearity and parallel nature with the wavelength axis, from which purity is established. In the case of the peak purity parameter the method is based on the absorbance weighted mean wavelength of a spectrum. Another method recently reported by Marr *et al.* [32] is multiple absorbance ratio correlation (MARC). It is based on the correlation between a reference spectrum from the data library and sample spectrum, and a comparison between these is based on regression analysis.

An additional derivative of the AR method is 'Spectral Suppression'. This is a spectral difference method, which is suitable for quantitation of two components. The method is based on finding the absorbance ratio at two suitable wavelengths for the compounds of interest. Following on from the equations above, Equation 1 can give a null relationship:

$$\Delta\Sigma_{1,2} = \Sigma_{\lambda_1} - K_{1,2} \cdot \Sigma_{\lambda_2} = 0 \tag{3}$$

The absorbance contribution of a single component at any concentration can be suppressed by applying the difference absorbance function, $A_{1,2}$

$$\Delta A_{1,2} = A_{\lambda_1} - K_{1,2} \cdot A_{\lambda_2} = 0 \tag{4}$$

## 114 UV detection

An interfering component with different spectral properties will then give a response (this can be positive or negative), where the magnitude is related to concentration [33]. An example of the operation of the method is given in Fig. 8.6, where the recommended British Pharmacopoeia (BP 1988) reversed-phase HPLC assay for the related impurity analysis of the $\beta$-blocker drug, atenolol, does not resolve two of the related impurities (PPA and diol). By application of the spectral suppression technique, where

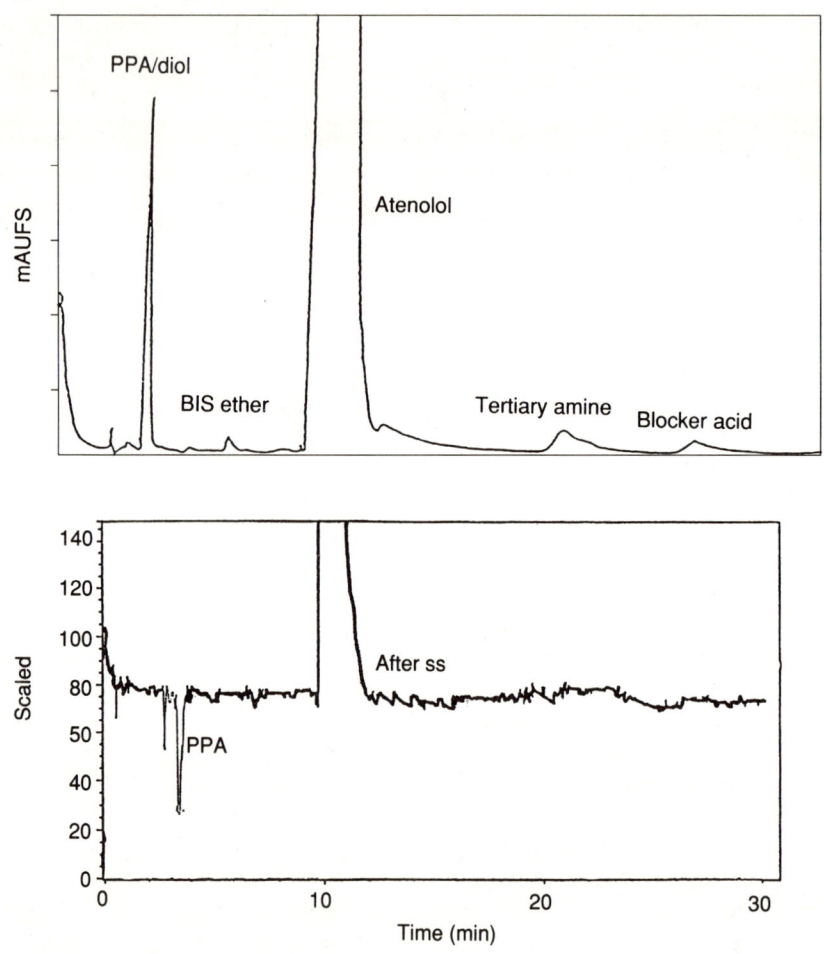

Fig. 8.6 *Atenolol and its related impurities separated by the BP method and the application of spectral suppression on the PPA/diol peak.*
*(a) The chromatogram of atenolol and related impurities, by the BP 1988 method, shows the overlap of the related impurities, PPA/diol. (b) The spectral suppression method is applied at $A = A_{226} - 5.12\, A_{280}$, which suppresses the diol and leaves the PPA as a negative peak, where amplitude is proportional to the concentration present.*

$$\Delta A_{PPA} = A_{226} - 5.12\, A_{280} \tag{5}$$

$$\Delta A_{diol} = A_{226} - 8.11\, A_{280} \tag{6}$$

then application of Equation 5 suppresses the diol and in Equation 6 the PPA, to give a measurable amplitude for the other component in turn [34, 35]. The software for this technique is available on commercial instruments and is able to detect a coeluting component down to 0.5% of the major component, if there is sufficient difference in the spectra of the two compounds. The technique is restricted to two components, but by a slight extension of the method to use molar extinction coefficients, multiple spectral suppression has been developed by Marr et al. [36].

In addition to these methods differentation in both the time domain and the less helpful mode of spectral derivative information can be useful to assess peak homogeneity [36], in a cognate manner to the method of spectral deconvolution in UV and visible spectrophotometry.

A novel development of the derivative information in both time and spectral domains, is the possibility of considering them in tandem in a three-dimensional spectroderivative chromatogram, as shown in Fig. 8.7. It is then possible to slice through this three-dimensional data matrix, to allow isolation of derivative chromatograms and this therefore leads to deconvolution of overlapping chromatographic peaks in the conventional chromatogram.

As has been indicated, the absorbance ratio, its derivatives and spectral suppression methods can be used for relatively swift assessment of peak purity. These methods represent a quick examination position and can be limited in their operation, requiring prior knowledge of at least one of the components when considering a binary system and as a rule all the components in a multicomponent coeluting system. In the case of the basic absorbance ratio methods, the detection level of a coeluting component is generally high (>10% w/w). But some improvement has resulted from methods such as multiple absorbance ratio correlation (MARC) where 1% w/w impurities have been picked up. As indicated above, these levels are also mirrored in the spectral suppression techniques and generally this level of detection is essential for trace impurity analysis. The method of choice relies purely on the expected level of impurity and the extent of prior knowledge and chromatographic overlap.

If little information on the possible coeluting components is available then the more involved the algorithms required and in these cases a move to the so-called chemometric methods may be essential. The basis of these methods is to estimate the maximum probable number of components in a chromatographic peak cluster, and the more advanced the method then the smaller the amount of prior information which must be known to operate the methods successfully. Methods such as 'iterative target transformation factor analysis' (ITTFA), which is one of the most powerful, will yield chromatograms and spectral information on the components present from the deconvolved data set. The multicomponent analysis methods above cover: spectral deconvolution by least squares or Fourier

116   UV detection

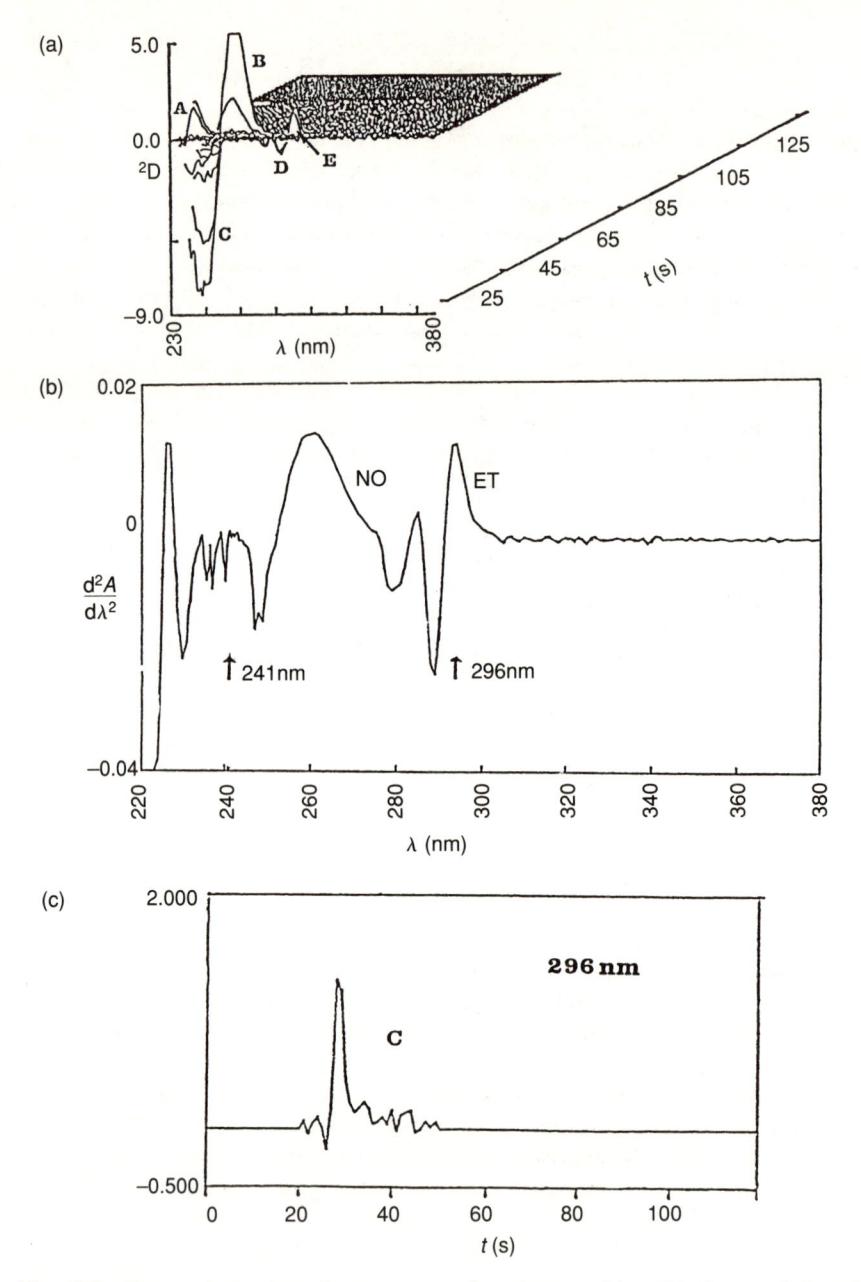

Fig. 8.7 *Spectroderivative chromatogram for the steroids ethinyloestradiol and noresthisterone.*
*Ethinyloestradiol and noresthisterone give coelution when chromatographed on a C18 column packing. By taking the derivative (time domain) chromatogram (a) of the three-dimensional data from the linear photodiode array detector and slicing through the data at 45s, the second derivative illustrating the unique area for the ethinyloestradiol at 296 nm is shown (b). From this it is possible to isolate the derivative in the time domain, the amplitude of which represents the concentration of ethinyloestradiol (c).*

transform; curve resolution (CR); curve fit; multiple regression analysis (MRA); principal component analysis (PCA); partial least squares (PLS) and iterative target transformation factor analysis (ITTFA).

The curve fitting routine of spectral deconvolution uses the wavelength data in a matrix-based least squares method [37]. In the method it is advantageous to know the sample spectra, although it is not essential. From the chromatographic peak, the component spectra concentration coefficients are computed alongside the wavelength data from the mixture spectra and this composite spectrum, which best fits the spectral data is generated. Repeats of the method at sequential time intervals throughout the elution profile yields a series of deconvolution chromatograms.

With the methods of MRA and PLS it is again best if the spectral information is available, due to the method relying on the similarity between the standard and actual spectra in the coeluting peak. In many of these methods overdetermined systems are operated, where the number of observation wavelengths at which data are recorded exceeds the number of possible components in the mixture. PLS is based on a curve fitting process between known and unknown component differences and these are minimized by taking the squares of the residuals. As in all these methods it is more helpful if certain restrictions are placed on the data to exclude negative values, linear range and noise. PLS is regularly operated in tandem with factor analysis (FA) where FA is initially utilized to reduce the large data matrix to a more manageable size. FA indicates the number of factors influencing the data set and the nature of physically significant parameters prior to using PLS. The data matrix is reduced to a minimum and this produces a series of orthogonal component vectors that envelop the significant information in the data set [38]. In addition, factor analysis has also been reported as an individual technique for interrogating chromatographic peaks. Notwithstanding the success of these methods in the research domain where MRA, PLS and FA have been shown to be powerful techniques for peak homogeneity testing, up to the present commercial software packages are not available for use with LDA detectors in HPLC.

Currently the commercial interest in this important field of chromatographic peak purity evaluation has been confined, although this is likely to change in the not too distant future. Of the commercial chemometrics packages for interrogation of chromatographic data, two of the earliest were developed by Kowalski and co-workers [39, 40] and are sold by Infometrix (Infometrix Inc, Washington, USA) for the Hewlett Packard systems. They are based on the principal components analysis method (PCA) and curve resolution. The PCA software is a statistically-based form of FA, which uses a number of assumptions and limitations about non-negative responses, linearity, noise, and minimal difference between the components. In PCA there is an extraction of the eigenvalues, in association with their respective eigenvectors, from a similarity matrix, resulting from different forms of the data, manipulated prior to analysis. In terms of the $(A, \lambda, t)$ data, it is represented by rows (spectra) and wavelengths (columns, elution profiles) to give the covariance matrix. The principal components from the factor analysis are abstract combinations of the chromatograms and spectra from the components. By introducing further data into the method which can be in the form of limitations on the data or further spectral information, this then allows transformation (rotation) of the abstract eigenvectors from which meaningful spectral and chromatographic data is extracted.

In the commercial package the PCA is limited to less than four components, where the success relies on the setting up of the method to extract the most useful data and the degree of overlap of the peaks.

The curve resolution method has the advantage over the PCA programme in that it operates on full spectrochromatographic data, so that little information about the components is required. However, it is useful to know some of the information. The method is basically a factor analysis method with considerable restrictions, and transforms abstract spectra or chromatograms from the factor analysis portion of the method into recognizable forms.

A third method is commercially available in the form of ITTFA from Phillips (Phillips Inc, The Netherlands). This is the most powerful package to date. The method deals with projections of a target test spectrum into hyperspace and compares this with the actual spectral data. The initial process is to rotate the abstract profiles to give targets and retesting of the targets until no further iteration is required. For quantitative information the identified components from the test spectrum can be utilized and the elution profiles are obtained by using a least squares best fit procedure. As with all these procedures there are certain restrictions concerning the chromatographic resolution and the degree of spectral similarity which will be tolerated.

Other methods for peak purity evaluation which have been introduced are not as powerful as the Phillips system and are generally incorporated in a more complete package for method development, peak purity and method validation; such as the ICOS (Hewlett Packard), Gold (Beckman) and Pesos (Perkin Elmer).

## 8.5 Conclusions

Research in the field of liquid chromatography is still progressing in both the areas of improved column selectivity and performance and in the area of enhanced detection sensitivity and selectivity. The universal detector appears not to be possible at present and the utilization of hybrid and specific detectors can apparently answer the requirements of the analyst in most areas. One of the prime developments in detector technology in the last decade has been the LDA system and there are likely to be many developments of this detector mode in the future. In terms of chromatographic peak purity and homogeneity validation, commercial developments have been somewhat slow in release. However the pace of development has speeded up in the last two years. It is anticipated that suitable software packages will soon be forthcoming and that the area will be opened up to include the introduction of more major developments of the expert system software which at present is only available through Phillips.

In contrast, for method development, optimization and method validation there are a number of software packages available which are generally very powerful. It is likely that in the next decade we will continue to see major steps forward in separation science and that the detector field will experience similar exciting developments, which will greatly assist the analytical chemist.

## References

1. Belchamber, R.M., Read, H., and Roberts, J.D (1986) in *Planer Chromatography*, Vol.1, (ed. R. Kaiser), Huthig Verlag, Heidelberg and Basel, pp. 207–20.
2. Nyarady, S., and Barkley, R. (1985) *Anal. Chem.*, **57**, 2074–79.
3. Karnes, H., and Farthing, D. (1988) *LC-GC*, **5**, 978–9.
4. Chien, C.F., Laub, R.J., Kopecni, M.M., and Smith, C.A. (1981) *J. Phys. Chem.*, **85**, 1864.
5. Wright, D.W. (1985) *Am. Lab.*, **17**, 74.
6. Liu, G. (1988) *J. Chromatogr.*, **441**, 372.
7. Yeung, E.S. (1986) *Anal. Chem.*, **89**, 1.
8. Pawliszyn, J. (1988) *ibid.*, **60**, 766.
9. Gauglitz, G. (1988) *ibid.*, **60**, 2609.
10. Ewing, A.G., and Wightman, R.M. (1984) *J. Neurochem.*, **43**, 570–75.
11. Johnson, D.C., and LaCourse, W.R. (1990) *Anal. Chem.*, **62**, 589A.
12. Radzik, D.M., and Lunte, S.M. (1989) *CRC Crit. Rev. Anal Chem.*, **20**, 317.
13. Small, H. (1983) *Anal. Chem.*, **55**, 235A.
14. Griffith, P.R., and Conroy, C.M. (1987) in *Advances in Chromatography*, Vol.34, (ed. J.C. Giddings), Marcel Dekker, New York, 384–92.
15. Helgeth, J.W., and Taylor, L.T. (1987) *Anal. Chem.*, **59**, 295.
16. Albert, K., and Bayer, E. (1988) *TrAc* **7**, 288.
17. Laude, D., and Wilkins, C.L. (1986) *TrAC* **5**, 60.
18. Hill, S., Ebdon, L., and Jones, P. (1986) *Anal. Proc.*, **23**, 6.
19. Ebdon, L., Wilkinson, J.R., and Jackson, K.W. (1982) *Anal. Chim. Acta*, **136**, 191.
20. Iroglic, K.J. (1987) *Mar. Chem.*, **22**, 265.
21. Smith, R.D., and Johnson, A.L. (1981) *Anal. Chem.*, **53**, 739.
22. Smith, R.D. (1989) *J.Phys. Chem.*, **93**, 5019.
23. Allen, M.H. (1989) *Rapid Commun. Mass Spectrom.*, **3**, 255.
24. Van Berkel, G.J., Glish, G.L., and McLuckey, S.A. (1990) *Anal. Chem.* **62**, 1284.
25. Vekey, K. (1989) *J. Chromatogr.*, **488**, 73.
26. Clark, B.J., Fell, A.F., and Jones, D.G. (1988) *J.Pharm. Biomed. Anal.*, **6**, 843.
27. Werzyn, J. (1990) *Anal. Chem.*, **62**, 1754.
28. Fell, A.F., and Clark, B.J. (1987) *Eur. Chromatogr. News*, **1**, 16.
29. Wright, A.G., Fell, A.F., and Berridge, J.C. (1987) *Chromatographia*, **24**, 533.
30. Fell, A.F., Clark, B.J., and Scott, H.P., (1984) *J. Chromatogr.*, **316**, 423.
31. Kostenko, V.G. (1986) *ibid.*, **355**, 296.
32. Marr, J.G.D., Seaton, G.G.R., Clark, B.J., and Fell, A.F. (1990) *ibid.*, **506**, 289.
33. Clark, B.J., Fell, A.F., and Westerlund, D. (1984) *ibid.*, **286**, 261.
34. Owino, E., Clark, B.J., and Fell, A.F. (1991) *J. Chromatogr. Sci.*, **29**, 298.
35. Owino, E., Clark, B.J., and Fell, A.F. (1991) *ibid.*, **29**, 450.
36. Marr, J.G.D., Horvath, P., Clark, B.J., and Fell, A.F. (1986) *Anal. Proc.*, **23**, 254.

37  Fasamade, A.A., Fell, A.F., and Scott, H.P. (1986) *Anal. Chim. Acta*, **187**, 233.
38  Otto, M., and Wegsheider, W. (1985) *Anal Chem*, **57**, 63.
39  Sharaf, M.A., and Kowalski, B.R. (1981) *ibid.*, **53**, 518.
40  Sanchez, E., Ramos, L.S., and Kowalski, B.R. (1987) *J. Chromatogr.*, **385**, 151.

# 9 Assessing the validity of spectroscopic information

## 9.1 Concepts of validation

The word 'validation' is much used and abused in analytical chemistry. It has become all things to all men. Sometimes it is used as an excuse for failure. How often have you heard utterances of the kind: 'But we ran the sample on a fully validated spectrometer'! Much confusion arises from misunderstandings of the nature of analytical processes and sources of variation.

For the purpose of this paper, I have taken the view that there are four distinct stages in the production of valid information, from the lowest level of data capture to the interpretation of results. These are:

Stage 1: Fundamental
Stage 2: System Control
Stage 3: Transformation
Stage 4: Interpretation

These stages are represented diagrammatically in Fig. 9.1. It is important to recognize that these exist in a strict hierarchy, and failure of integrity at a lower level invalidates and precludes progress to the next level up. Remember that useful computer acronym, GIGO: Garbage in; Garbage out.

The quality of information extracted via a series of processes is only a good as the inherent data quality and presumes that the data gathered is relevant to the information sought.

## 9.2 Stage 1: Fundamental

The first stage, which I have called fundamental, is comprised of two parts: the integrity of the data and the integrity of the sample. It is the former which is normally referred to in terms of validation. Here we are concerned with the physics associated with the spectrometer and the stability of its electronic systems.

Challenges to the system involving standard filters for wavelength and absorbance accuracy enable the optimum operating range to be assessed and monitored. Statements on accuracy and precision, short term lamp noise and

122    *Assessing spectroscopic information*

```
┌─────────────┬─────────────┬─────────────┬─────────────┐
│  STAGE 1    │  STAGE 2    │  STAGE 3    │  STAGE 4    │
│ Fundamental │ System Control │ Transformation │ Interpretation │
└─────────────┴─────────────┴─────────────┴─────────────┘
```

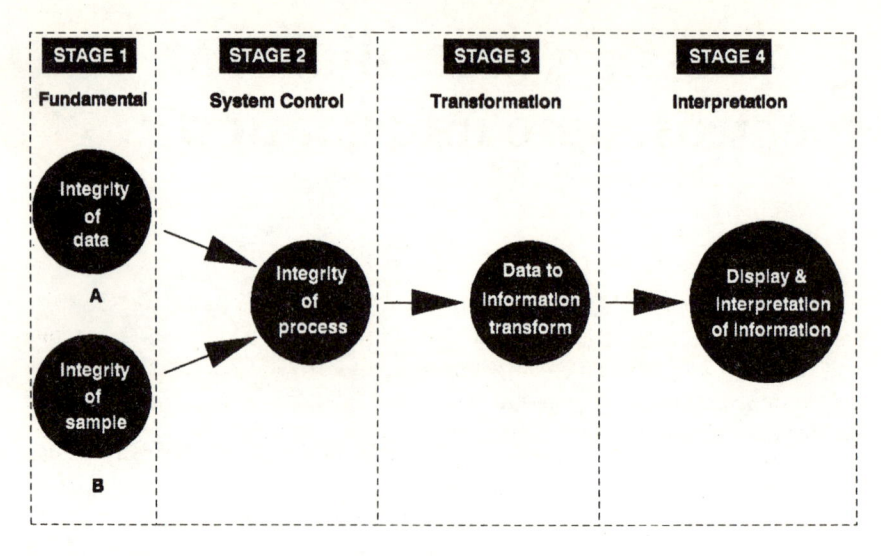

Fig. 9.1  *Assessing the validity of spectroscopic information.*

drift, signal-to-noise ratio etc., belong in this category. Over longer periods of time, control chart approaches allow the assessment of the maintenance of the instrument and long term ageing of components.

Vital though these measurements are, far too often they are seen as validating the method rather than just the first stage in the process.

The second aspect of the fundamental stage is usually neglected entirely or assumed to be of little importance. The integrity of the sample, be it a true sample or standard reference material, is probably the greatest challenge to any spectroscopic method.

Matters to be addressed here include:

(i)    the quality of the solvents (chemical and particulate);
(ii)   ensuring absence of bias in the standard employed;
(iii)  ensuring that the sample is indeed representative of the batch;
(iv)   preparing the sample and solution in such a way as to minimize error. For example, gravimetric as opposed to volumetric dilution or automated procedures; and
(v)    optimizing the sample preparation to enable the instrument to operate within its optimal dynamic range.

The use of liquid standards for monitoring the accuracy and precision of the spectrometer goes some way to challenging the spectroscopic system as a whole.

Once the analytical scientist is satisfied with these aspects of integrity it is then possible to move on to Stage 2.

## 9.3  Stage 2: System control

Here I am using the word control in a broader sense than is customary. In its usual sense it refers to the operability of analytical system. Control of such factors as static or flow presentation of the sample to the incident beam of radiation, the cleanliness and handling of cuvettes, consideration of temperature and sensitivity of the sample to pH or ionic strength are critical to the acquisition of good data.

However, the broader aspect of control refers to the requirement of 'good laboratory practice' (GLP), 'good manufacturing practice' (GMP) and BS5250. In essence these requirements are a common sense approaches to the capture and recording of analytical data. Indeed, the documentation of methods, results, calibration records, and standardization procedures are part of the control mechanism of the greatest source of variance; namely, the operator. The discipline imposed by these codes of practice is designed to ensure that the principle of 'Say what you Do' and 'Do what you Say' is observed.

While introducing these concepts at Stage 2, the same principles apply to succeeding stages. It is only when that confidence and a degree of assurance of the underlying data quality have been gained that the next step should be undertaken.

## 9.4  Stage 3: Transformation

The data information transform will only produce valid information if the data quality is adequate and if the data is relevant. It may seem trivial to state this but it is remarkable how often one sees people attempting to extract information from data which isn't there!

One of the keys to success is to ensure that the parameter space is wide enough and that the experimental design is geared to providing data embracing

```
            DATA ANALYSIS
                USING
             STATISTICAL
               METHODS
           ↙              ↘
    INFERENTIAL          DESCRIPTIVE
  i.e building models   i.e. detection and
     for predicting        interpretation
       behaviour            of patterns

  e.g Least squares    e.g. multivariate data
       methods            analysis methods
```

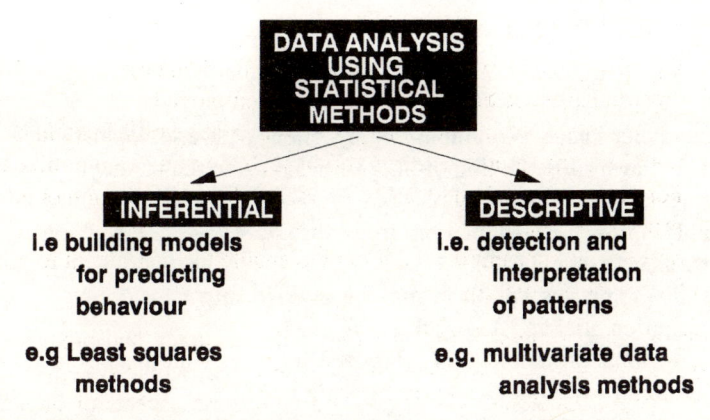

Fig. 9.2  *The two groups of statistical methods.*

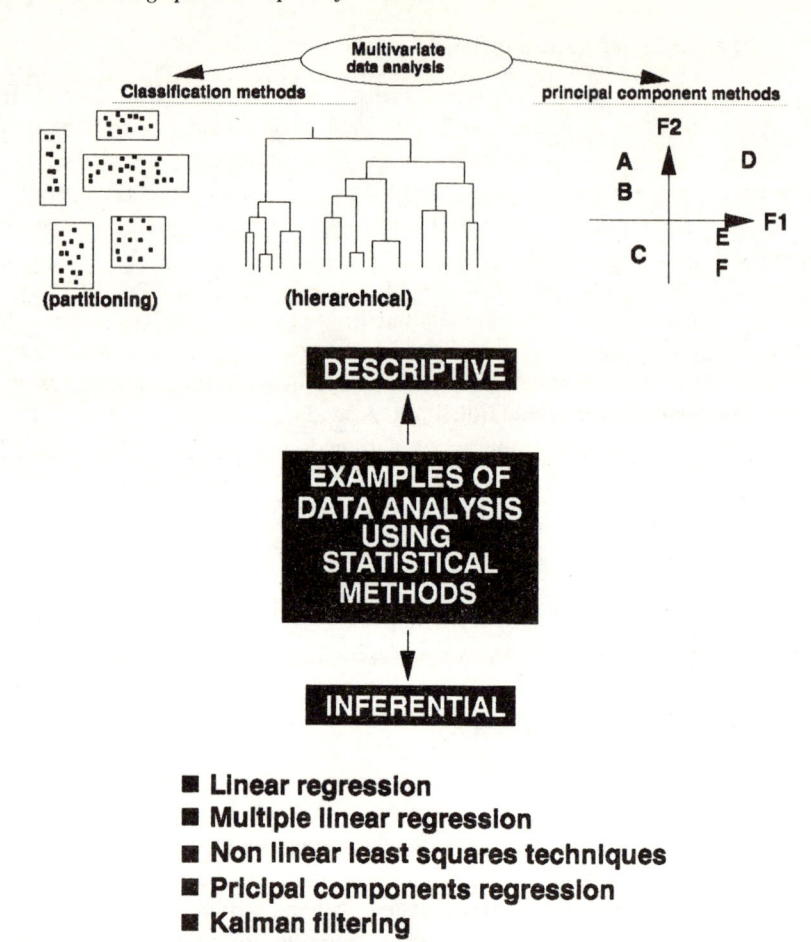

Fig. 9.3 *Examples of data analysis.*

this parameter space. No amount of excellent data and sophisticated mathematics will avail if the region of interest lies outside the parameter space of the experiment.

Quality must be designed in from the start and not rely on statistical procedures to rescue it.

The chemometrics approach may be divided into two groups of statistical methods: inferential and descriptive (Fig. 9.2).

Prior to transformation, it is often best to look at the data with descriptive methods in an exploratory way using such techniques as principal component analysis, clustering etc., to assess the dimensionality and shape of the data set (Fig. 9.3).

Once satisfied from an initial look, the inferential methodologies based on model building can be applied. These may range from the ubiquitous least squares multicomponent analyses, through principle component regression to partial least square techniques, to name but a few.

It is essential, however, to understand the underlying assumptions and pitfalls before applying the technique. One of the great modern problems is the consummate ease with which elegant but inappropriate methods can be applied to multivariate data sets. For valid information to be derived from valid data assumes the choosing of an appropriate transform. The principle 'keep it simple' is to be recommended.

## 9.5  Stage 4: Interpretation

The obtaining of valid information from spectroscopic data is not usually the end of the matter. The interpretation of the information is required. Beyond this is usually a neglected area for the technically inclined; namely, once you have valid answers you must market them usually to a non-technical audience.

This must involve display of information in ways digestible to the non-specialist. Remember: the end user for the information knows not, and probably cares less, about Stages 1 to 3. These for him are the given.

Fortunately, graphical techniques can enable a seemingly insurmountable barrier to be climbed between the analytical scientist and the accountant or general manager.

## 9.6  Summary

In conclusion, I have sought to convince you that validation is not a single process but a series of stages, each dependent on the integrity of the previous stage. It is broader than just instrumental standardization, as it embraces all the regulatory aspects of documentation and control.

It is the professional responsibility of the analytical scientist to ensure that all stages have been properly thought through.

I have deliberately used the term 'analytical scientist' as the journey towards valid information transcends analytical chemistry alone.

# 10 Practical experiments

## 10.1 Introduction

The experiments in this chapter are designed to be carried out under the supervision of an experienced analytical chemist in a suitably equipped laboratory. The experiments can be run on almost any spectrophotometer. Instrumental settings are for guidance only.

Care should be exercised when preparing and handling the solutions, especially in the use of acids and solutions of potassium dichromate.

We hope these experiments will be of use as training exercises in both academic and industrial laboratories. The questions are designed as prompts for discussions between the supervising analyst and the student or trainee.

Further discussion of all the topics in the practical experiments can be found in the third volume of the series: *Techniques in Visible and Ultraviolet Spectrometry. Practical Absorption Spectrometr*y (edited by A. Knowles and C. Burgess), Chapman & Hall (1984).

## 10.2 The experiments

The experiments have been divided into four convenient sessions:

Session 1     **Fundamental principles**
Instrumental parameters
Absorbance and transmittance
Calibration

Session 2     **Collection of good quality spectra**
Sample preparation
Effects of temperature, pH and solvent

Session 3     **Quantitative aspects**
Two component assay
Derivative spectroscopy

Session 4     **Multicomponent calculations**
Multiple linear regression
Kalman filter regression
Principle components regression and factor analysis

# Session 1
# Fundamental principles

**Objectives of the experiment**
1. To demonstrate how instrumental settings influence spectra.
2. To demonstrate ways of calibrating spectrophotometers.
3. To introduce the concepts of absorbance and transmittance.

**Outline**

These experiments are designed to complement Chapters 1 and 2. In the first experiment the spectrum of a holmium filter is used to demonstrate the effect of slit width and scan speed. The spectrum is also compared with a spectrum of holmium perchlorate solution.

In the second experiment the data from solutions containing a range of concentrations of potassium dichromate is used to assess linearity and the behaviour of these parameters with concentration.

**Experimental details**

(a) *Recording the spectrum of a holmium filter*

Select the following instrumental conditions or the nearest equivalent.

Mode: absorbance
Slit width: 1 nm
Scan speed: 120 nm min$^{-1}$
Wavelength range: 200–600 nm

Ensure the instrument gives a zero scan with the cell holders empty.

Place a holmium filter in the sample compartment and scan the filter with the reference holder empty. Obtain a printout of the spectrum and record the wavelength and absorbance of the maxima at about 280, 360 and 460 nm.

Scan and print out the region 400–500 nm using the smallest slit width the instrument will allow and the slowest scan speed. Repeat the scan with the fastest scan speed and widest slit width. Finally, scan the region with the fastest scan speed and smallest slit width. Typical scans are given in Fig. 10.1

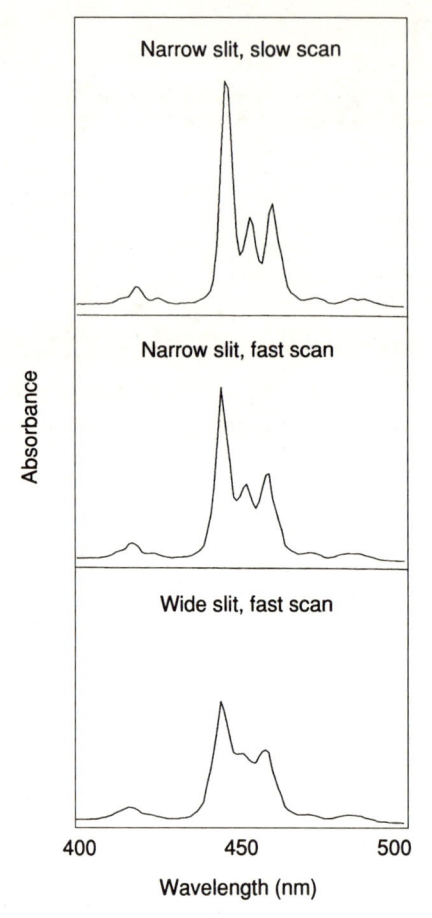

Fig. 10.1 *Spectra of a holmium filter demonstrating the effect of varying the slit width and scan speed.*

If you are using a diode array instrument you will not be able to vary these parameters, but record a spectrum and compare the results with those from a scanning instrument shown in Fig. 10.1.

Find the wavelength of maximum absorbance of the band near 360 nm and measure the absorbance at the maximum and 1 nm either side of the maximum.

Remove the holmium oxide filter and replace with a cell filled with holmium perchlorate solution. Scan and print out the spectrum of the solution. Compare the spectrum of the solution with the spectrum from the filter.

## (b) The concepts of absorbance and transmittance

Transfer 60 mg of potassium dichromate accurately weighed to a 100 ml volumetric flask. Dissolve in and dilute to volume, with 0.05 M sulphuric acid. Dilute this solution with 0.05 M sulphuric acid as follows:

| Solution | Dilution | Concentration |
|---|---|---|
| A | 10 to 100 | |
| B | 5 to 100 | |
| C | 2 to 100 | |
| D | 10 to 50 | |
| E | 10 to 25 | |
| F | 15 to 25 | |
| G | 2 to 200 | |
| H | 2 to 500 | |

Record the absorbance of each solution against the 0.05 M sulphuric acid at 257 nm and 350 nm in a 1 cm cell.

Measure the absorbance of solution D at 349 nm and 351 nm.

## Calculations

Calculate the concentration for each solution.

Plot the following graphs using the data at each wavelength:

1. Absorbance against concentration
2. Transmittance against concentration
3. Transmittance against log concentration
4. Molar absorptivity against log concentration

Typical plots are shown in Figs. 10.2 to 10.5 inclusive.

## Questions

### Calibration

1. Compare the effect on the absorbance values when readings are taken ±1 nm either side of the maximum absorbance value for potassium dichromate and the holmium filter. How well could each of these spectra be used for calibration of spectrophotometers? Which scale would they be suitable for calibrating?

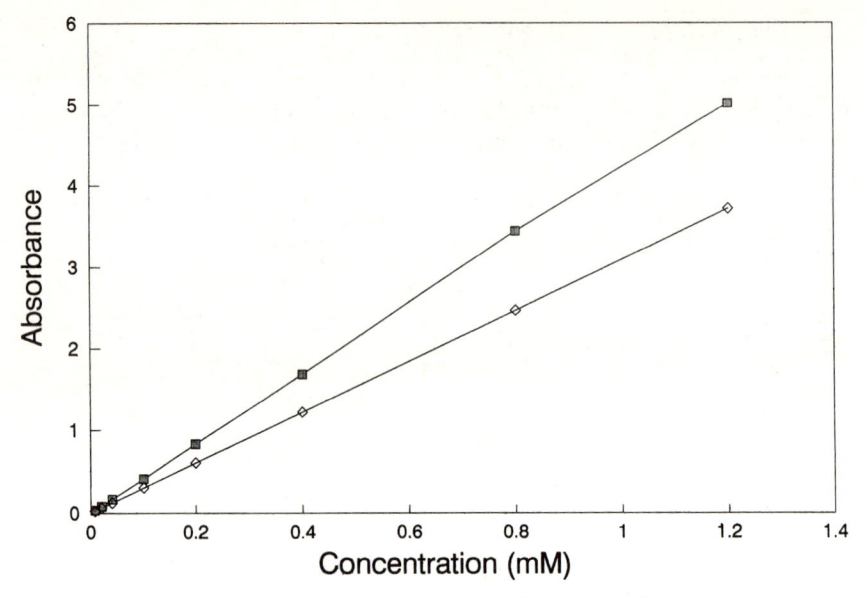

Fig. 10.2 *Absorbance against concentration at (■) 257nm and (◊) 350nm.*

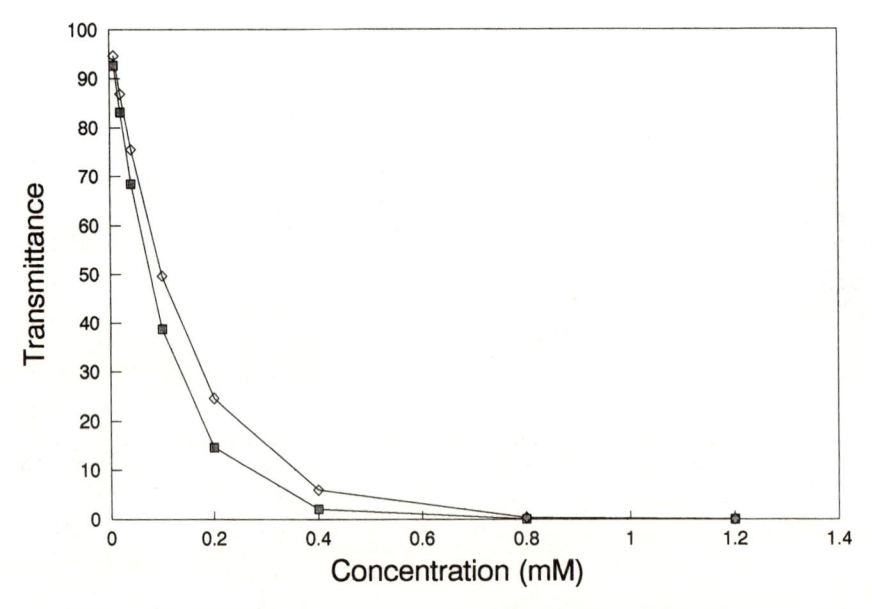

Fig. 10.3 *Transmittance against concentration at (■) 257nm and (◊) 350nm.*

Fundamental principles 131

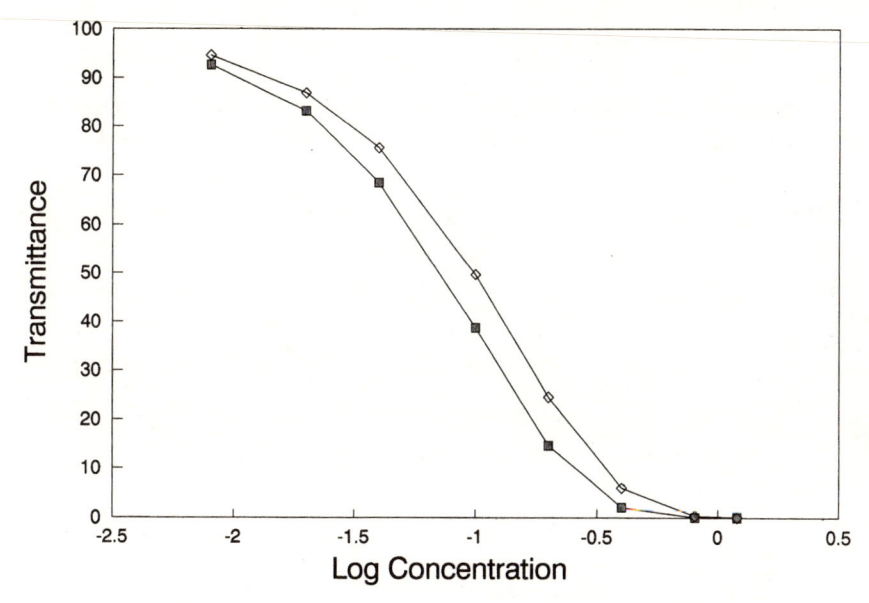

Fig. 10.4  *Transmittance against log concentration at (■) 257nm and (◊) 350nm.*

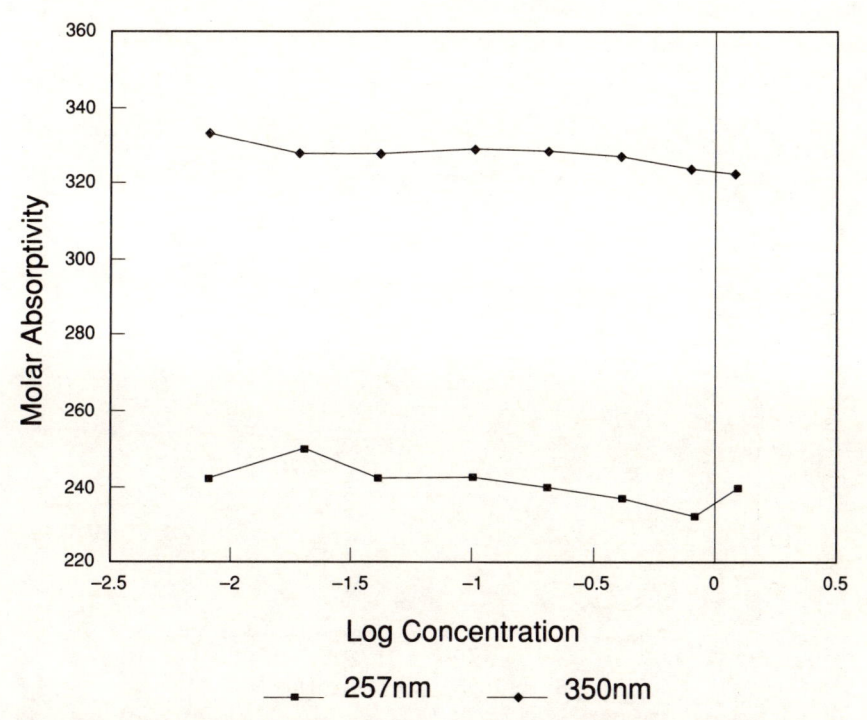

—■— 257nm   —♦— 350nm

Fig. 10.5  *Molar absorptivity against log concentration at (■) 257nm and (◊) 350nm.*

## Absorbance and transmittance

1. Look at the four plots from potassium dichromate. Which parameter shows a linear response with concentration and is therefore suitable for single point calibration?

2. What factors influence linearity and which plots bring out the optimum linear range best?

# Session 2
# Factors that influence the collection of good quality spectra

## Objectives of the experiment
1. To demonstrate the factors of sample preparation that influence the measurement of good quality spectra.
2. To demonstrate the effects of temperature, pH and solvent on the repeatability of spectra.

## Outline
Eight solutions of methyl orange are prepared, each in a different pH buffer, and the spectra of the solutions are recorded. The spectra are plotted and examined for suitability and the estimated measurement errors are examined for any unusual features.

Up to ten wavelengths are selected by the operator and the absorbance and standard deviations (if available) at these wavelengths are recorded. This data is used in the practical sessions on day 4.

## Experimental details
1.  BUFFERS

Prepare eight McIlvaine's buffer solutions spanning the pH range 2.2 to 5.2 as described below. These solutions have a near constant ionic strength.

(a)  *0.2 M disodium hydrogen orthophosphate*

Weigh 28.4 g of disodium hydrogen orthophosphate into a beaker and dissolve in 100 ml distilled water. Transfer the contents of the beaker to a one litre flask and make up to volume with distilled water.

### (b) 0.1 M citric acid

Weigh 21.0 g of citric acid into a beaker and dissolve in 100 ml of distilled water. Transfer the contents of the beaker to a one litre flask and dilute to volume with distilled water.

### (c) Buffer solutions

Using appropriate volumetric glassware prepare eight buffer solutions by mixing aliquots of solutions 1(a) and 1(b) as follows:

| Approximate pH | Volume of $Na_2HPO_4$ | Volume of citric acid |
|---|---|---|
| 2.0 | 1.0 | 49.0 |
| 2.6 | 5.5 | 44.5 |
| 3.0 | 10.0 | 40.0 |
| 3.4 | 13.0 | 37.0 |
| 3.8 | 18.0 | 32.0 |
| 4.2 | 21.0 | 29.0 |
| 4.6 | 23.0 | 27.0 |
| 5.2 | 27.0 | 23.0 |

Divide each mixture into two equal portions of 25 ml.

## 2. METHYL ORANGE SOLUTIONS

### (a) Stock solution

Weigh 0.5 g methyl orange into a beaker and dissolve it in distilled water. Transfer the solution to a one litre flask and dilute to volume with distilled water.

### (b) Buffered solutions

Add 0.5 ml of methyl orange stock solution (2a) to one portion of each of the buffer solutions (1c). These are the analytical solutions and the other portion of the buffer is the blank.

### (c) Non-buffered solutions

Add 0.5 ml to methyl orange stock solution (2a) to 25 ml of distilled water contained in a volumetric flask.

Repeat the above procedure using ethanol, methanol, and acetone as the solvents.

## 3. SPECTRAL MEASUREMENT

### (a) *Effect of pH*

Clean a 10 mm pathlength cuvette and fill it with one of the blank buffer solutions, insert the cuvette in the instrument and measure the BALANCE or background or run a background correction as necessary for the instrument you are using, over the range 250–650 nm. Empty the cuvette and wash it with the corresponding analytical solution. Measure the spectrum over the same range and store the spectrum in the instrument or on disc if possible.

Repeat the sequence for each blank buffer and corresponding analytical solution.

### (b) *Effect of temperature*

Investigate the effect of temperature on the spectrum of the pH 3.4 methyl orange buffer solution in the range about 10 to 40 °C. Accurate temperatures and temperature control are not necessary for this experiment and it is necessary to measure the spectra at only three temperatures.

### (c) *Influence of solvent*

Record the spectra (200–600 nm) of the non-buffered solutions using the appropriate solvent as the blank.

## 4. CALCULATIONS

### (a) *Spectral plots*

Plot all eight spectra run at different pHs (3(a)) on the same sheet of paper so that they use the same wavelength and absorbance scales. The spectra should all have zero absorbance at about 620 nm. Identify the isosbestic points and comment on their significance. If the isosbestic point at about 470 nm is not clear then the spectra are not good enough for the calculations in session 4.

### (b) *Further calculations on the data*

Select any ten wavelengths between 300 and 600 nm and tabulate the absorbance values of each spectrum at these wavelengths. Keep this data for later use.

Plot the three spectra recorded at different temperatures (3(b)) on one sheet of paper. Comment on the spectral changes that occur and suggest reasons for these changes.

Overlay the spectra recorded with the non-buffered solutions (3(c)). Comment on the significance of the spectral shifts that are observed. How do these shifts relate to the nature of the solvents?

# Session 3
# Quantitative aspects

**Objectives of the experiment**

1. To demonstrate how simple mixtures can be assayed using simultaneous equations.
2. To demonstrate the use of derivative spectroscopy.

**Outline**

Standard spectra of tyrosine and phenylalanine are recorded and used to quantify the concentrations of these two compounds in a mixture using data at two wavelengths. A third component, tryptophan, can also be quantified using data at three wavelengths.

The derivative spectra of the simple solutions and mixtures are examined to see which compounds can be quantified by derivative spectroscopy.

**Experimental details**

*Preparation of the standard solutions*

Transfer 50 mg of phenylalanine accurately weighed, to a 100 ml volumetric flask. Dissolve and make up to volume with 0.1 M hydrochloric acid. This is the phenylalanine standard.

Transfer 50 mg of tyrosine accurately weighed, to a 100 ml flask. Dissolve in and dilute to volume with 0.1 M hydrochloric acid. Dilute this solution 10 ml to 100 ml with 0.1 M hydrochloric acid. This is the tyrosine standard solution.

*Preparation of the test solution*

A test solution should be prepared in 0.1 M hydrochloric acid which gives a spectrum in the same concentration range as the standards.

Record the spectra and measure the absorbance at 256 nm and 275 nm of the two standard solutions and the first test solution provided and record the results.

A third amino acid to give a UV spectrum is tryptophan. Prepare a standard solution of tryptophan as follows: transfer 50 mg accurately weighed, to a 100 ml volumetric flask, and dissolve in and dilute to volume with 0.1 M hydrochloric acid. Dilute this solution 10 ml to 100 ml with 0.1 M hydrochloric acid. This is the standard solution.

Record and overlay the spectra of all three standards. Choose a wavelength in addition to 256 nm and 275 nm where the three spectra show significant differences.

Record the absorbance of each standard and the second test solution at 256 nm, 275 nm and the third wavelength.

Typical spectra for the three amino acids are shown in Fig 10.6. Note how converting the phenylalanine spectrum to digital data has reduced the fine structure.

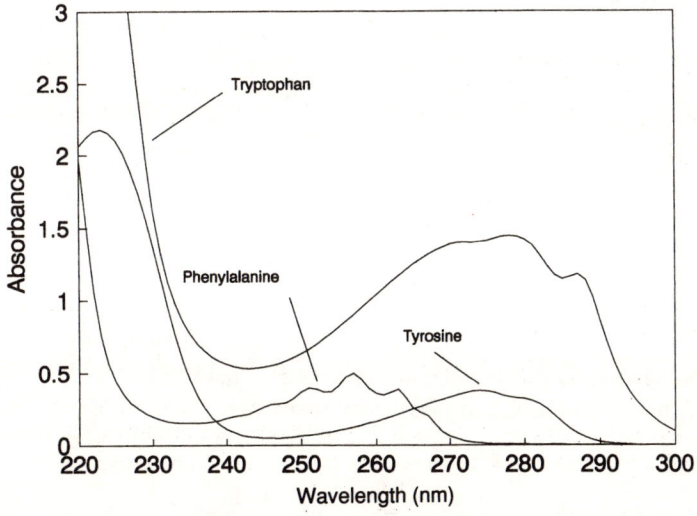

Fig. 10.6 *Spectra of phenylalanine, tyrosine, and tryptophan.*

## Calculations

The absorbance of a mixture of compounds is the sum of the individual absorbance values of each compound provided the compounds do not interact. The simplest way of calculating the concentration of mixtures is to use simultaneous equations. Thus

$$\text{Absorbance} = [\text{tyr}]E_{\text{tyr}} + [\text{phe}]E_{\text{phe}}$$

where $E$ is the molar absorptivity of the respective amino acid.

Calculate the concentration of phenylalanine and tyrosine in the test solution.

Construct simultaneous equations for the two standards and the test solution at each wavelength and solve for the concentration of each amino acid in the first test solution.

The incorporation of a third component, tryptophan, gives rise to a further unknown, and three simultaneous equations are necessary to calculate the molar absorptivities.

Construct simultaneous equations and calculate the concentrations of each amino acid in the second test solution using the data at 256 nm, 275 nm and the third wavelength chosen.

## Derivative spectroscopy

Derivative spectroscopy can be used to enhance fine structure and eliminate broad peaks.

Examine the first, second and fourth derivative spectra of each solution if they can be obtained from the spectrometer you are using. Using the second derivative spectra, select wavelengths that you feel are appropriate and construct simultaneous equations to allow you to calculate the concentrations of amino acids in the test mixtures.

Examine the effect on the signal-to-noise of recording and averaging several second derivative spectra.

Examine the effect of changing the number of data points use to calculate the derivative spectrum.

## Questions

1. Which derivative spectra are most useful for quantitation?
2. Which amino acids can be assayed by derivative spectroscopy and why?

# Session 4
# Multicomponent calculations using spectral data

## Objectives of the experiments
1. To demonstrate the measurement requirements for multicomponent assays.
2. To demonstrate alternative methods of carrying out quantitative multicomponent calculations.
3. To demonstrate the principles of factor analysis.

## Outline
The set of methyl orange spectra at different pHs which were collected in session 2, will be used as the data for the calculations. No new experiments are needed.

Up to ten wavelengths in the spectra are selected by the operator and the absorbances at these wavelengths are recorded for each buffered methyl orange spectra.

A simple linear regression model is assumed and the two extreme pH solutions are defined as the two standard solutions. The concentrations of the two species in the remaining solutions are then calculated using linear regression methods.

If time allows this experiment can be extended very simply in two ways:

1. The absorbance spectra can be converted into first or second derivative spectra and the calculations repeated with these.
2. A third component can be added to the solutions, either at a constant concentration or at differing concentrations, and the calculations repeated to investigate the effect of a contaminant.

## Calculations
These calculations should be read in conjunction with the chapter on multicomponent analysis (Chapter 4).

### (a) Simple linear regression

Use the methyl orange pH data collected in Session 2. In the calculations you may wish to use fewer than the ten wavelengths; however do take the opportunity of exploring the effect of varying the number of wavelengths selected and the actual wavelengths selected. These calculations can be carried out quite easily using a calculator. Do not round off the numbers during the intermediate calculation stages. (Why?).

Refer to standard text books on matrix algebra[1] if you are unsure about these topics.

The solutions at pH 2.2 and 5.2 are defined as the reference solutions of the two forms of methyl orange. Create the matrix $E$ of the absorbance values. For example for six wavelengths:

$$E = \begin{vmatrix} 0.0649 & 0.4542 \\ 0.3312 & 0.6808 \\ 1.0080 & 0.6716 \\ 1.1900 & 0.2270 \\ 0.2570 & 0.0184 \\ 0.0061 & 0.0020 \end{vmatrix} \quad \updownarrow \text{wavelength}$$

with pH along the top.

Calculate the matrices $E'E$ and $(E'E)^{-1}$ where $E'$ is the transpose of $E$

$$E'E = \begin{vmatrix} 2.612156 & 1.206802 \\ 1.206802 & 1.172704 \end{vmatrix}$$

$$(E'E)^{-1} = \begin{vmatrix} 0.7297864 & -0.7510059 \\ -0.7510059 & 1.6255722 \end{vmatrix}$$

Select one of the remaining six solutions and create a matrix $A$ for the same set of wavelengths used for $E$, for example:

$$A = \begin{vmatrix} 0.1026 \\ 0.3640 \\ 0.9965 \\ 1.0880 \\ 0.2315 \\ 0.0061 \end{vmatrix}$$

Calculate the matrices $E'A$ and $B = (E'E)^{-1}(E'A)$

$$E'A = \begin{vmatrix} 2.45570 \\ 1.19476 \end{vmatrix}$$

$$B = (E'E)^{-1}(E'A) = \begin{vmatrix} 0.89486 \\ 0.09792 \end{vmatrix}$$

The elements of the final matrix, $B$, are the proportions of the two methyl orange forms in the solution you have chosen. The sum of the elements should be 1.00.

Estimates of the errors in these values are obtained as follows. Use the estimated amounts of the components ($B$) to calculate the estimated absorbance of the solution A.

$$A1 = EB = \begin{vmatrix} 0.10255 \\ 0.36305 \\ 0.96779 \\ 1.087117 \\ 0.231782 \\ 0.005655 \end{vmatrix}$$

and hence the residuals from the measured values $R = A - A1$

$$R = \begin{vmatrix} 0.00005 \\ 0.00095 \\ -0.00129 \\ 0.00088 \\ -0.00028 \\ 0.00045 \end{vmatrix}$$

In the ideal case, all of the elements of $R$ should be zero.

Estimate the residual variance $S^2$ from

$$S^2 = \frac{R'R}{n-2}$$

$$S^2 = 9.078503 \times 10^{-7}$$

where $n$ is number of wavelengths chosen.

Finally, calculate the covariance matrix, $V = (E'E)^{-1} \cdot S^2$

$$V = \begin{vmatrix} 6.6254 \times 10^{-7} & -6.8180 \times 10^{-7} \\ -6.8180 \times 10^{-7} & 1.4758 \times 10^{-6} \end{vmatrix}$$

The diagonal elements of $V$ are the variance estimates of the proportions of each component. The square roots of these are standard error estimates.

(b)  *Kalman filter.*

An outline of the Kalman filter method is given in Chapter 4 on multicomponent analysis. Although the calculations involved are very simple there are many of them and it is not practical to carry them out manually. If you have access to a spectrophotometer with the calculations built in, use the methyl orange solutions to examine the filter. In order to use the routine, the spectra of the solutions should have been measured and stored on the instrument. The instrument manual should give clear instructions on how to use the method. Again define the solutions at pH 2.2 and 5.2 as standards and use the intermediate pH solutions as the test spectra.

(c)  *Simple factor analysis*

Explanatory notes about factor analysis are given in Chapter 4 on multicomponent analysis.

There are a few software packages available commercially for carrying out principal component analysis and principle component regression. One popular package is 'Unscrambler'[2]. If you have access to suitable software, select any ten wavelengths and note the absorbance values of the analytical solutions at these wavelengths. Feed the data into the software and find the eigenvalues and eigenvectors.

Identify the significant ones from the printed tests of significance. As there are only two components in the mixture you can expect to see two significant factors. Is this the case?

If you have software that will carry out principal component regression, use one spectrum as a test and the remainder as a training set. You will need to enter concentration values for the two components in each of the spectra in the training set.

## References

1   Bronson, R (1970) *Matrix methods — An Introduction*, Academic Press, London.
2   Unscrambler — CAMO, A/S Jarleveien 4, N-7041 Trondheim, Norway.

# Index

Absolute matching  69
Absorbance  9
  experiment  129–32
  index  113
  standards  36–7, 42–7, 65
Absorbance ratio-ing (AR)  112–13, 115
Absorptiometric multichannel detection  107–18
Absorption of light
  common bands  79–84
  laws of  6–9
Absorption spectrometry instrumentation  17–23
Accuracy  12–14
Adaptive Kalman filter  60
Aliphatic compounds  79
Alkyl groups  79–80
Amperometric detectors  101–2
Aromatic compounds  82–4
Atomic absorption (AA)  104–6
Autoscaling  69
Auxochromes  79

Beam splitters  22–3
Beer–Lambert law  9, 26, 53–4, 59
  maximum scaling  68
  plot  12, 14
Benzene
  absorption bands  82–3, 84
  slit width effects  11
  vapour phase and solution spectra  7
Blazed gratings  21
Boyle, Robert  2
Buffers  133–4

Calculation models  59
Calibration
  experiments  129
  multicomponent  60–1
  multivariate techniques  53
  spectral libraries  64–5
Capillary columns  98, 106
Carotenoids  80–1
Cells  23
Chance ON10 filters  42–3
Charge-coupled devices (CCD)  107
Charge transfer effects  84
Chemical validity of data  64

Chemiluminescence detectors  98
Chiral compounds, separation of  98–9
Chromatographic peak purity and homogeneity
  validation  112–18
Chromophores  79–82
CIE Tristimulus colour space  90–3
Circular dichroism (CD)  99
Colour  1–2, 88–96
  chip systems  90
  rules  84
Colour-play effects  95
Comparison techniques  69–77
Conductivity detectors  103
Contour diagrams  33–4
Correction of spectra  32–4
Correlation
  spectral  73, 74–7
  spectra-structure  78–87
Coulochem Electrode Array System  102
Coulometric detectors  101–2
Curve resolution (CR)  117, 118
Cut-off filters  50, 65
Cyanine dyes  81
Cytochrome C  11
Czerny–Turner monochromators  21, 22

Data
  normalization  67–8
  storage  65–6
  validity  121–2
Databases, spectroscopic  63–4
Deconvolution method  24
Derivative information  115, 116
Derivative spectroscopy  24, 136, 138
Descriptive methods  123, 124
Detectors  23, 99–110
Deuterium  37–8
  arc lamps  18, 21
Didymium filters  38–9
Diode array spectrophotometers  52, 55
Dipole effects  83
Double beam spectrometers  17, 20
Double monochromators  21
Dynamic quenching  28–9

Ebert monochromators  21, 22
Effective spectral bandwidth  12
Electrochemical detection  101–2

Electrosprays (ES) 106
Emission spectra 33–4
Enantiometric compounds 98–9
Energy transfer
    fluorescence 27–8
    molecular luminescence 26
Enzyme assays 24
Ethanol 48, 49
Ethene 3, 4
Euclidian metric 72–3, 74–6
Excitation-emission matrix (EEM) 33
Excitation monochromators 32–3
Excitation spectra 32–4
Experiments 126–42
Extinction value 9

Factor analysis (FA) 117–18, 142
Feature selection 69–72
Fieser's rules 81
Filters
    cut-off 50, 65
    glass 36–40, 42–4, 50
    Kalman 58–60, 142
    liquid 36–7, 40–2, 44–7, 50
Fixed wavelength detectors 100
Flame photometric detectors 98
Flavonoid pigments 82
Fluorescence 24, 25–6
    intensity 26–9
    polorization 28, 29–30
Fluorimetry
    instrumentation 17, 30–2
    sensitivity 24
Fourier transform infrared (FTIR) 104
Frequency 3–5
Full spectral matching 72

Gas liquid chromatography (GLC) 98–9
Glass filters 36–40, 42–4, 50
Grating monochromators 21, 32

High Performance Liquid Chromatography (HPLC) 99–119
    developments 52
    luminescence spectroscopy 24
Holmium filters 38, 127–8
Holmium perchlorate solution 40, 128
Holographic gratings 21
Hues 91–4
Hybrid detectors 98, 103–7

Identity of materials 64
Immunoassays 24
Indicators, screening of 95
Inferential methods 123, 125
Inner filter effects 27–8, 32
Inorganic compounds 79
Instrumentation 17–34

Intensity 93
Interpretation of information 125
Ion chromatography 103
Iterative target transformation factor analysis (ITTFA) 115, 118

JCAMP-DX transfer protocol 66, 67

Kalman filter 58–60, 142

Labelled data records (LDR) 66
$L^*a^*b^*$ representation 93–4
Lambert's law 6–9
Lasers 103
Least squares regression 54–8
Libraries, spectral 63–77, 112
Light screening detectors 103
Light sources 18, 21, 31–2
Linear photodiode array (LDA) detectors 107–12
Linear regression 55–8, 62, 140–2
Lippert mechanism 28
Liquid chromatographic detection 97–120
Liquid filters 36–7, 40–2, 44–7, 50
Littrow monochromators 21, 22
Luminescence multichannel systems 107
Luminescence spectroscopy 23–6
$L^*u^*v^*$ representation 93

McCrone wavelength standard 40–1
Mahalanobis distances 70–2
Maleic acid 48–9
Manhattan metric 72–3, 74–6
Mass spectrometry (MS) 106–7
Maximum scaling 68
Measurement
    colour 90–5
    models 59
Mercury lamps 21, 31
Methyl orange 55–8, 133–4
Microscopy 24
Molar absorptivity 9
Molecular luminescence 24–6
Monochromators 21–2, 32
Multichannel detectors 107–18
Multicomponent analysis 52–62, 139–42
    liquid chromatographic detection 97–120
Multifunctional detectors 103
Multiple absorbance ratio correlation (MARC) 113, 115
Multiple linear regression 62
Multiple regression analysis (MRA) 117
Multivariate calibration techniques 53
Multiwavelength data 97

NBS glass filters 42–3
NBS liquid filters 46, 47
Normalization, data 67–8

# Index

NPL filters   44
Nuclear magnetic resonance (NMR)   104

Partial least squares (PLS)   117
Particle concept   2
Peak matching   69–70
Peak purity and homogeneity validation   112–18
Peak purity parameter   113
Peak tracking   110–12
pH
  experiments   135
  fluorescence   28
Phosphorescence   26
Phosphorimetry   24
Photodecomposition   26
  fluorimetry   31
Photoionization detectors   98
Photoluminescence   23
Photomultipliers   23
Photon concept   2–3
Polarimetric detection   99
Polarization, flourescence   28, 29–30
Polyaromatic compounds   82–4
Positive matching   69
Potassium chloride solution   50
Potassium dichromate solution   44–6
Precision   12–14
Preparation of samples   64
Principal components analysis (PCA)   61–2, 117–18, 142

Quantum counters   31
Quantum yields   26–8
Quartz-halogen lamps   18, 21, 31
Quenching phenomena   28–9

Radiant energy   1–3
Raman scattering   24, 25, 34
Range scaling   68
Rapid scanning spectroscopic detection   107–18
Rayleigh scattering   24, 25
  contour spectrum   33
  fluorimetry   31
  polarization   30
Reflectivity   22–3
Refraction   5
Refractive index   101
Residual spectra   55
Reverse optics   17, 19

Samarium perchlorate solution   40–1
Samples
  preparation   64
  validity   122
Saturation   91–2
Scaling, spectral   68–9

Schott NG-4 filters   42–4
Search techniques   66–8
Self-scanned detection   109–10
Similarity metric   67
Simple least squares regression   54–8, 140–2
Single beam spectrometers   17, 18
Slit width   10–12, 13
Solid standards   36–40, 42–4, 50
Solution spectra   6, 7
  normalization   68
Solution standards   36–7, 40–2, 44–7, 50
Solvents, fluorescence   28, 29
Spectra
  fluorescence   32–4
  origin and nature   6, 7, 78–9
  quality   10–12
  valid   14–15
Spectral bandwidth   10–12
Spectral correlation   73, 74–7
Spectral libraries   63–77, 112
Spectral scaling   68–9
Spectral suppression   113–15
Spectra-structure correlation   78–87
Spectroderivative chromatograms   115, 116
Spectrophotometers, diode array   52, 55
Spectrophotometric measurements   1–16
Spectroscopic databases   63–4
Spectroscopic validity   64–5
Stand-alone detectors   109–10
Standard deviation   14
Standards   15, 35–51
  see also Calibration
State models   59
Static quenching   28–9
Stern–Volmer equation   28
Stokes shift   26
Stray light   10, 12
  spectral libraries   65
  standards   36–7, 47–51
Synchronous spectra   34
Synthetic spectra   55
System control   123

Tautomeric groups   84
Temperature
  control   23, 32
  experiments   135
  fluorescence   28
Thermosprays   106
Thin Layer Chromatography (TLC)   97–9
  luminescence spectroscopy   24
Thomson's solution   46–7
Transformation   123–5
Transmittance   9, 129–32
Tristimulus colour space   90–3
Tungsten filament lamps   18

Universal detectors   101

146  *Index*

Validity
  chromatographic peak purity and
    homogeneity 112–18
  spectral libraries  63–5
  spectroscopic information  121–5
Vapour phase spectra  6, 7
Variable single spectra  34
Variable wavelength detectors  100–1
Velocity  3–5
Viscosity  28
Visible radiation  1–3
Voltammetric detectors  101–2
Vycor filters  50

Water  49
Wavelength  3–5
  High Performance Liquid Chromatography  52
  standards  36, 37–41, 64–5
Wavenumber  5
Wave-particle duality  2
Weighted regression  55
Woodward rules  81

Xenon lamps  21, 31–2

Zechmeister's rules  81